Ground-Penetrating Radar

An Introduction

for Archaeologists

Any sufficiently advanced technology is
indistinguishable from magic.

Arthur C. Clarke

Ground-Penetrating Radar

An Introduction
for Archaeologists

Lawrence B. Conyers

&

Dean Goodman

A Division of
ROWMAN & LITTLEFIELD PUBLISHERS, INC.
Walnut Creek • Lanham • New York • Oxford

ALTAMIRA PRESS
A Division of Rowman & Littlefield Publishers, Inc.
1630 North Main Street, #367
Walnut Creek, CA 94596
www.altamirapress.com

Rowman & Littlefield Publishers, Inc.
4720 Boston Way
Lanham, MD 20706

12 Hid's Copse Road
Cumnor Hill, Oxford OX2 9JJ, England

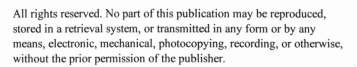

Copyright © 1997 by AltaMira Press

Production and Editorial Services: David Featherstone
Editorial Management: Joanna Ebenstein
Cover Design: Joanna Ebenstein

British Library Cataloguing in Publication Information Available

Library of Congress Cataloging-in-Publication Data

Conyers, Lawrence B.
    Ground penetrating radar : an introduction for archaeologists / Lawrence B. Conyers
and Dean Goodman.
        p.   cm.
    Includes index.
    ISBN 0-7619-8927-7 (cloth.) — ISBN 0-7619-8928-5 (pbk.)
    1. Geophysics in archaeology. 2. Ground penetrating radar.
    I. Goodman, Dean, 1958-   . II. Title.
    CC79.G46C66 1997
    930.1'028—dc21   97-4828
                                                                       CIP

Printed in the United States of America

♾™ The paper used in this publication meets the minimum requirements of American
National Standard for Information Sciences—Permanence of Paper for Printed Library
Materials, ANSI/NISO Z39.48–1992.

Contents

9588   D(CON)

About the Authors   7

Acknowledgments   9

Chapter 1   Introduction   **11**
Geophysical Methods Used in Archaeology   14
Ground-Penetrating Radar Methods   15
The History of Ground-Penetrating Radar in Archaeology   18

Chapter 2   The GPR Method   **23**
Generation, Propagation, and Reflection of Radar Energy   27
Production of Continuous Reflection Profiles   29
Data Recording   31
Physical Parameters Affecting Radar Transmission   31
Radar Propagation   35
Antenna Frequency Constraints   40
Focusing and Scattering Effects   52
Signal Attenuation   53
The Near-Field Effect   55

Chapter 3   GPR Equipment and Data Gathering   **57**
Description of GPR Equipment   57
Recording Data in the Field   60
Orientation of Surface Transects   62
Acquisition of Survey Data   64
Equipment and Software Adjustments   68

Chapter 4   Post-Acquisition Data Processing   **77**
Background-Removing Filters   78
F-k Filters   80
Deconvolution   81
Migration   82

Chapter 5	**Synthetic GPR Modeling**	**83**
	Creating a Synthetic Radargram	86
	Synthetic Radargram Applications in Archaeology	91
	Synthetic Radargrams Compared to GPR Profiles	94
Chapter 6	**Time-Depth Analyses**	**107**
	Reflected-Wave Methods	109
	Direct-Wave Methods	119
	Laboratory Measurements	130
	Velocity Analyses Conclusions	133
Chapter 7	**The Use of GPR Data to Map Buried Surfaces and**	
	Archaeological Features	**137**
	Ancient Landscape Reconstruction	137
	Data Interpretation	138
	Buried Structure Identification	141
	Paleotopographic Maps	142
	Ancient Drainage Patterns and Topography	142
	Computer-Generated Three-Dimensional Maps	145
Chapter 8	**Amplitude Analysis in GPR Studies**	**149**
	Amplitude Slice Maps on Level Ground	156
	Amplitude Time Slices on Uneven Ground	163
	Three-Dimensional Mapping	165
	Identification of Features Invisible in Two-Dimensional	
	Profiles	168
	Horizon-Slice Maps	172
	Integration of GPR Amplitude Data with Resistivity and	
	Magnetic Maps	178
	The Use of Amplitude Time Slices to Search for Vertical	
	Features	184
	The Use of Amplitude Time Slices to Image Features in	
	the Near-Field Zone	189
Chapter 9	**Conclusions**	**195**
References Cited		207
Index		225
Index of Authors Cited		231
Color Plates following page		94

About the Authors

Lawrence B. Conyers received a bachelor of science degree in geology from Oregon State University and a master of science degree from Arizona State University, also in geology. He holds both a M.A. degree and a Ph.D. in anthropology from the University of Colorado, Boulder. Conyers worked extensively in petroleum exploration before turning his attention to the use of geophysical techniques to map near-surface geological and archaeological features. He presently teaches in the Department of Anthropology at the University of Denver and conducts geophysical research at archaeological sites throughout the United States and Central America.

Dean Goodman received a bachelor of science in applied geophysics from the University of California, Los Angeles, and a master of science degree in marine geophysics from Oregon State University. His Ph.D., in applied marine physics, is from the University of Miami. Since 1989, Goodman has headed the Geophysical Archaeometry Laboratory of the University of Miami Japan Division. His primary research has been developing new GPR imaging techniques at archaeological sites in Japan. He is also a guest researcher at the Nara National Cultural Properties Research Institute in Japan.

Acknowledgments

Our greatest thanks go to the many people who have helped in our research in and field development of ground-penetrating radar techniques over the years. Much appreciation goes to Jeff Lucius and Mike Powers, of the U.S. Geological Survey, and to Gary Olhoeft, of the Colorado School of Mines, for their many hours of technical assistance and encouragement. Payson Sheets, of the University of Colorado, was a constant source of motivation during many trying times in both the field and the laboratory. Thanks also go to the many experts in GPR who gave us their time to review and comment on our earlier drafts of the manuscript. These thanks especially go to Floyd McCoy, Clark Dobbs, Marilyn Beaudry, and Bruce Bevan. For help in Japan, we would like to thank Yasushi Nishimura, of the Nara National Cultural Properties Research Institite, as well as Katsumi Arita, Hiromichi Hongo, Hideo Sakai, Takao Uno, Yoshinori Hosoguchi, and Yoshiko Ouyachi. Special thanks go to Koji Tobita for his dedicated field assistance in Japan. Acknowledgment for their help in logistics and field assistance is also due to Rinsaku Yamamoto, Noboru Tsujiguchi, Tokuo Yamamoto, J. Daniel Rogers, Bruce Smith, James Price, Mark Lynott, Ohkita Masaaki, Ishii Katsumi, Yukio Maehata, and the Matsue City Archaeological Department.

Chapter 1

Introduction

In today's climate of rescue archaeology, cultural resource management, and the prevalent ethic of site conservation, non-invasive methods of subsurface analysis are becoming increasingly important. With excavation budgets severely restricted, and strict political and resource conservation considerations that must be considered, in many cases it is no longer feasible or desirable to excavate large areas. Archaeological excavation strategies have therefore changed dramatically in the last decade for much of the archaeological community. At many sites, the expense and time necessary to carry out large-scale excavations preclude the gathering of extensive information about buried cultural resources not readily visible on the surface. Many times it is not feasible to excavate at all, which severely hampers the archaeologist who is only familiar with traditional methods of gathering data. Remote sensing methods, including geophysical surveys, must therefore be employed that can gather important subsurface information without time-consuming and costly digging.

Recent advances in geophysical exploration and subsurface delineation techniques have proven to aid greatly in site identification and mapping, as well as in expediting future excavation strategies. Ground-penetrating radar (GPR) is one of these methods that has recently gained a wide acceptance in the archaeological community. This method can quickly and accurately define many buried archaeological features and important stratigraphy in three dimensions, saving both time and money.

All archaeological studies rely for the most part on data that are patchy and of uneven quality. Often there is much arbitrariness and accident in the discovery of archaeological remains, and subsequent excavation strategies are many times based on little knowledge about the extent or distribution of artifacts. Rarely are sites of any size wholly excavated, even in the unusual cases where budgets are large and field programs cover many years. In many cases, a wide amount of terrain must be covered in a superficial fashion, and detailed excavations using standard excavation techniques are either widely spaced or must be carried out quickly during rescue operations. When survey and excavation budgets are limited, only small portions of sites can actually be excavated; and information about the extent and nature of the remaining portions must by necessity be extrapolated from a very limited data set. In a worst-case scenario, construction operations or other human disturbance may destroy undiscovered or unexcavated portions of a site prior to any archaeological study. Due to its ease and excellent resolution, GPR can be used in these cases to identify sensitive archaeological areas that can be avoided and preserved.

Geophysical data can often tell many things about an archaeological site that cannot be learned any other way. At deeply buried sites, ground-penetrating radar data can be just as meaningful as artifacts or other excavated features. It is thus essential that archaeologists integrate geophysical data into their overall conceptual framework when interpreting a site. Utilized in this way, geophysics is not just a tool for discovering buried archaeological materials and features, but becomes a part of the overall assemblage of a site.

Ground-penetrating-radar surveys allow archaeologists to cover a wide area in a short period of time, with excellent subsurface resolution of buried archaeological features and their related stratigraphy. When soil and sediment conditions are suitable, some radar systems have been able to resolve stratigraphy and other features at depths in excess of forty meters (Davis and Annan 1992). More typically, GPR is used to map features of archaeological interest at depths from a few tens of centimeters to five meters in depth. Radar surveys can not only identify buried features for possible future excavation but also interpolate between excavations, projecting archaeological knowledge into areas that have not yet been, or may never be, excavated.

Ground-penetrating-radar information is acquired by reflecting radar waves off subsurface features in a way that is similar to how radar methods are used to detect airplanes in the sky. The radar waves are propagated in distinct pulses from a surface antenna; reflected off buried objects, features or bedding contacts; and detected back at the source by a receiving antenna. As radar pulses are transmitted through various layers on their way to the buried target feature, their velocity changes depending on the electrical and magnetic properties of the material through which they are traveling. When the travel times of the energy pulses are measured and their velocity through the ground is known, distance (or depth in the ground) can be accurately measured. In the GPR method, radar antennas are moved along the ground and two-dimensional profiles of a large number of periodic reflections are created, producing a profile of subsurface stratigraphy and archaeological features along transects (Fig. 1). When data are acquired in a series of transects within a grid and the reflections are correlated and processed, an accurate three-dimensional picture of buried features and associated stratigraphy can be constructed.

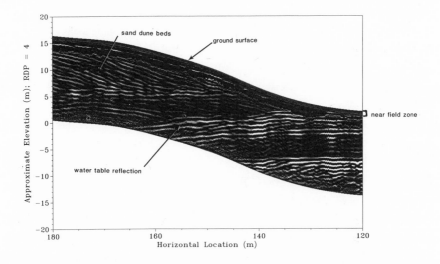

Figure 1. Representative GPR profile across a sand dune at the Great Sand Dunes National Monument, Colorado. The black and white lines represent subsurface reflections from bedding planes. (Adapted from Olhoeft 1994.)

GEOPHYSICAL METHODS USED IN ARCHAEOLOGY

There are a number of other geophysical methods that have also been applied with varying success to archaeological exploration. Resistivity, electromagnetic induction, and magnetic susceptibility are used most often by archaeologists (Clark 1996). Resistivity surveys are carried out by placing four electrodes on the ground surface, inducing an electrical current—which is nothing more than the displacement of electrons through a conductive medium—between two of them, and measuring the difference in voltage between the other two. The electrical conductivity (and its inverse, resistivity) of the soil, sediment, and possible buried archaeological features can then be measured. Depending on a number of factors, the most important being the distance between the electrodes on the ground surface, an estimate of the depth in the ground to which the current passes can be made. Because different soils and buried features have a varying capacity to conduct an electrical current, changes in the electrical resistivity across the ground can indicate the presence of buried features. When a series of measurements is made in a grid and the results are plotted on a map, anomalous areas that may represent features of interest are apparent.

In a similar fashion, electromagnetic induction (EM) has also been widely used in geophysical archaeological exploration. Electromagnetic energy is the simultaneous coupling of an electrical and a magnetic field. These fields are created perpendicular to one another and feed on each other as they project out from a source. Electromagnetic fields typically occur when an electrical current polarizes the material through which it is propagating and creates a subsidiary magnetic field. Once the field is created, it does not propagate by electron displacement as in an electrical current, but by the reactions of billions of atoms within nearby material to the induced field.

The EM geophysical mapping method induces what is called a primary electromagnetic field from one source, which is held above the ground. As this field passes into the earth, it creates a secondary field within the materials that comprise its sphere of influence. A second sensing device, held at the surface some distance away from the primary source, then measures this secondary field. More conductive materials will dissipate the EM field, while resistive materials transmit more of the electromagnetic energy. As the EM tool is moved along surface transects, measurements of the conductivity within the ground are obtained. When the results of many transects within a grid are plotted and contoured in a fashion similar to resistivity data, anomalies can be identified that may be related to conductivity changes in soil types or within large archaeological features.

While both the resistivity and EM methods have been used successfully in archaeological exploration, they can only crudely calculate the depth of anomalies that are discovered because the path of the energy that is transmitted (in resistivity surveys) or the dimensions of the sphere of influence in the ground (in EM surveys) can only be estimated. Their benefit is that they many times can be used in wet or clay-rich areas where GPR reflection techniques are sometimes less successful.

Magnetic susceptibility surveys measure minute changes in the earth's magnetic field that can be affected by changes in the magnetism of near-surface features. In archaeological contexts, these changes can be caused by hearths, baked clay floors of pit houses, buried kilns, and many other anthropogenic modifications to an ancient landscape. Magnetic surveys, like resistivity and EM, have been successfully applied in areas where GPR is a less appropriate method; but as with the other methods, the depth of any anomalies discovered can only be estimated.

GROUND-PENETRATING RADAR METHODS

Ground-penetrating-radar units have recently become very portable, and complete systems can be easily transported in a few backpacks into remote areas. Most systems can be powered from any high-amperage battery such as a 12-volt car battery, from portable electrical generators, or directly from 110-volt alternating currents. With some recently developed GPR units, a few small rechargeable batteries can power all of the radar equipment and computers necessary for data acquisition and field processing for many hours.

Some of the earliest-model GPR systems recorded raw subsurface reflection data on paper printouts that allowed little post-acquisition processing. Although these radar systems, which are still in use, can many times yield valuable subsurface information, modern digital systems record reflection data on a computer hard drive, digital tape, or floppy disks. Most personal computers can then process, filter, and enhance these raw data once the researcher is back in the office. Computer enhancement and manipulation software programs have recently allowed for rapid advancements in data enhancement and interpretation, leading to much-improved subsurface resolution. Equipment miniaturization has recently allowed computer processing of the acquired radar data to occur immediately, and interpretation can sometimes begin while still in the field.

The success of GPR surveys in archaeology is to a great extent dependent on soil and sediment mineralogy, clay content, ground moisture, depth of burial, surface topography, and vegetation. GPR is not a geophysical method that can be arbitrarily applied to any geographic or archaeological setting, although with thoughtful modifications to acquisition and data-processing methodology, it can be adapted to many differing site conditions. In the past it has been assumed that GPR surveys would only be successful in areas where soils and underlying sediment are extremely dry and nonconductive (Vickers and Dolphin 1975). Although radar-wave penetration and the ability to reflect energy back to the surface are enhanced in a dry environment, recent research that will be discussed in this book has demonstrated that dryness is not necessarily a prerequisite for GPR surveys. Modern methods of computer enhancement and processing have proven that meaningful data can be obtained, sometimes even in very wet ground conditions.

Proposed radar surveys must be analyzed in advance with respect to the proper equipment, field methods, and acquisition parameters. In order for a survey to be successful, it is often necessary to adapt survey techniques to take into account the geographic and geologic setting of the surveys, with adjustments for the size and depth of the archaeological features and stratigraphy to be studied.

Once GPR data have been acquired in the field and recorded digitally on a computer, there are a wide range of data-processing and interpretation techniques available. Depending on the archaeological questions to be asked and the quality of the radar reflection data acquired, these processing techniques can also be varied and modified to meet specific needs.

This book discusses the basic theoretical aspects of the GPR method in archaeology and some of the field and post-acquisition techniques used to acquire, process, and interpret the data. Some of the more complicated aspects of the GPR method, such as detailed electromagnetic theory, complicated equations used in data processing, and details about the schematic components of the equipment, are not immediately applicable to general archaeological investigations; information on these is held to a bare minimum. For the average archaeological user, these questions can be answered by a radar expert or electrical engineer, or researched in greater detail from the references cited in the text. Most archaeologists only want to know whether the method will work for them, how to go about doing a survey, and most importantly, how to process and interpret the data after acquisition. These are the aspects of GPR data acquisition and analysis that are emphasized in this book.

The field of archaeology is slowly making the transition from a "softer" science that relies only on data acquired from standard excavation techniques to a "harder" science that involves many chemical, physical, geological, and geophysical techniques. While this transition may seem difficult to many, it is a progression that must take place. It is our goal to provide a basic understanding of one of the more successful geophysical mapping methods so that archaeologists will no longer regard archaeological geophysics as the use of "spooky black boxes" and indecipherable equations that require the knowledge of outside consultants who have a physical science background. Ground-penetrating radar is not a method only for geophysicists who perform some kind of "magic" in the field. Most archaeologists trained today have more than enough scientific background to allow them to understand and use this exciting and promising method of archaeological mapping. All it takes is some field experience—the background that will allow prudent acquisition procedures—a determination to try GPR out, and the patience to process and interpret the data once they are acquired.

This book contains examples of GPR surveys and results from a number of different archaeological sites around the world. In each case study presented, differing GPR acquisition, processing, and interpretation techniques were applied to sites that had differing soil and overburden conditions, depths of burial, and size and character of archaeological features. The resulting reflection data were processed using a variety of techniques to create useful maps of the buried archaeology. Each case study provides an illustration of a successful methodology, illustrating both what was done right and some of the pitfalls that can occur. Some of the sites discussed in this book are:

- The Ceren site in El Salvador, a Mayan village buried in volcanic ash, where twenty-two buried houses and the prehistoric landscape were mapped by GPR in three dimensions.
- The Nyutabaru burial mound in Japan, where a burial chamber surrounded by a moat was discovered and later excavated.
- The Suzu kiln site in Japan, where GPR was successful in finding three buried ceramic kilns located along a hillside.
- The Shawnee Creek site in Missouri, where pit dwellings and other archaeological features were identified and later confirmed by excavation.

It is hoped that some of the techniques used in these and other case studies presented can be readily applied to many other archaeological sites with equal success.

THE HISTORY OF GROUND-PENETRATING RADAR IN ARCHAEOLOGY

Ground-penetrating radar was initially developed as a geophysical prospecting technique to locate buried objects or cavities such as pipes, tunnels, and mine shafts (Fullagar and Livleybrooks 1994). The GPR method has also been used to define lithologic contacts (Basson et al. 1994; Jol and Smith 1992; van Heteren et al. 1994), faults (Deng et al. 1994), and bedding planes and joint systems in rocks (Bjelm 1980; Cook 1973, 1975; Dolphin et al. 1974; Moffatt and Puskar 1976). Ground-penetrating-radar technology has also been used to investigate buried soil units (Collins 1992; Doolittle 1982; Doolittle and Asmussen 1992; Johnson et al. 1980; Olson and Doolittle 1985; Shih and Doolittle 1984) and the depth to groundwater (Beres and Haeni 1991; Doolittle and Asmussen 1992; van Overmeeren 1994).

The archaeological community was quick to grasp the potential of using GPR to locate and help define buried archaeological features and associated stratigraphic units. One of the first applications to archaeology was conducted at Chaco Canyon, New Mexico (Vickers et al. 1976). The purpose of this 1975 study was to discover the location of possible buried walls at depths of up to one meter. A number of experimental traverses were made at four different sites, and the interpretation of the paper field records concluded that a few of the anomalous radar reflections recorded on some of the profiles represented the location of buried walls.

The studies at Chaco Canyon were followed by a number of GPR applications in historical archaeology. Radar surveys were used to search for buried barn walls, stone walls, and underground storage cellars in many different areas, primarily in the eastern United States (Bevan and Kenyon 1975; Kenyon 1977). In these early studies, what were described as "radar echoes" were recognized from buried walls, and depth estimates were made using approximate velocity measurements estimated from local soil characteristics.

These initial successes in historical archaeological applications were followed in 1979 at the Hala Sultan Tekke site in Cyprus (Fischer et al. 1980) and the Ceren site in El Salvador (Sheets et al. 1985). Both GPR surveys produced unprocessed reflection profiles containing anomalies that were used to delineate buried walls, house platforms, and other archaeological features. In both cases, the material that buried the archaeological sites was extremely dry and therefore almost "transparent" to radar energy propagation, making the reflection records relatively uncomplicated to interpret.

During 1982 and 1983, a GPR survey was carried out at the Red Bay, Labrador, archaeological site in Canada in an attempt to locate graves, buried artifacts, and house walls associated with a sixteenth-century Basque whaling village (Vaughan 1986). This area was a challenging test for GPR mapping because the soils were wet and the overburden contained large cobbles and other natural features that had the potential to obscure radar reflection data. Artifacts and archaeological features that were buried by up to two meters of beach deposits and peat were discovered in many of the GPR printouts acquired at Red Bay. Some velocity tests that calculate the radar travel rate were carried out in order to convert radar travel time to an approximate depth in the ground. Archaeological excavations later tested the origins of the GPR anomalies identified. It was determined that grave goods, consisting of bone and metal artifacts, did not contrast enough with the surrounding beach deposits to appear as distinct anomalies, but the disturbed soil in some graves did appear as anomalous zones on radar profiles. Numerous large cobbles in the overlying material were found to have produced some of the anomalous zones, somewhat complicating data interpretation. Other anomalies were found to have been generated from buried walls that consisted of piles of these same beach cobbles.

A comprehensive series of GPR surveys was conducted in Japan in the mid-1980s in order to locate buried sixth-century A.D. houses, burial mounds and what were called "cultural layers" (Imai et al. 1987). In these studies, radar surveys proved capable of identifying ancient dwellings with sunken clay floors that were buried by as much as two meters of volcanic pumice and loamy soil. The interface of the clay house floors with the overlying pumice produced distinctive reflections that were easily recognizable on GPR profiles. After the GPR data were interpreted and archaeological anomalies delineated, portions of one of the sites were excavated. The locations of radar anomalies were then compared to the locations of houses, burial mounds, and associated trenches, with excellent correlation. Three cultural layers, occurring in buried soil horizons containing stone artifacts associated with different periods of occupation, were recognized on some GPR profiles.

Throughout the late 1980s and early 1990s, GPR continued to be used successfully in a number of archaeological investigations. In most cases these studies were "anomaly hunting" exercises that were used to search for possible cultural features at unknown or estimated depths that could later be excavated.

Radar studies became more numerous in archaeological research in the late 1980s and were primarily employed to locate buried features for later excavation or for cultural resource management and preservation. The Fort Laramie Historic Site in Wyoming (De Vore 1990), Roman walls at York, England (Stove and Addyman 1989), the Rockwell Mount site in Illinois and the Kualoa Park on Oahu, Hawaii, (Doolittle and Miller 1991) are some examples of this kind of work.

Prior to 1993, the most encompassing and successful archaeological application of GPR was that employed in the mapping of the houses and burial mounds in Japan, discussed above (Imai et al. 1987). This success was followed by numerous additional GPR surveys in Japan, conducted by Goodman and his colleagues (Goodman 1994; Goodman and Nishimura 1993; Goodman et al. 1994; Goodman et al. 1995). These studies pioneered the use of many GPR acquisition and data-processing techniques, some of which will be discussed in this book. Time- and depth-slice maps, computer simulated two-dimensional models, and three-dimensional reconstructions of buried features were all employed to discover and map buried ceramic kilns, burial mounds surrounded by moats, and individual stone-lined graves. A wide range of burial conditions were encountered, and in some cases these were studied in computer models prior to data acquisition in order to determine the antennas that would work best for local conditions. These models were also used as an aid in interpretation during data analysis.

One of the major GPR advancements for archaeology was the realization that radar reflections, measured in time, could be defined in real depth when radar wave velocity was determined (Imai et al. 1987; Vaughan 1986). The identification of reflections that correspond to horizons of archaeological interest was also used in a limited way to map related stratigraphy and buried topography (Conyers 1995b; Imai et al. 1987). Recently the application of two-dimensional computer simulation and three-dimensional processing techniques (Goodman et al. 1994, 1995) has shown that even radar data that does not yield immediately visible reflections can still contain valuable reflection data when further analyzed by computer. In the future, the use of these and other new techniques will greatly expand the utility of GPR exploration and mapping in archaeology.

Many archaeologists who employ GPR at their sites are concerned only with identifying buried anomalies that represent features of interest (see, for example, Butler et al. 1994; Sternberg and McGill 1995; Tyson 1994). Although this type of GPR application is valuable in that buried features can be identified immediately, this book will illustrate how the radar reflection data acquired in these types of studies can be further enhanced by a number of computer processing, interpretation, and display techniques. With little additional effort, computer technology allows for the construction of maps that can be interpreted in ways that will yield much more information about a site than was previously thought.

GPR's ability to not only map buried structures and other cultural features non-invasively and in real depth, but also to reconstruct the ancient landscape of a site, will become increasingly important. Computer enhancement of raw GPR reflection data will also become widespread as researchers increase their familiarity with some of the computer processing techniques discussed in this book and many others that are also presently available.

Chapter 2

The GPR Method

The GPR method involves the transmission of high-frequency electromagnetic radio (radar) pulses into the earth and measuring the time elapsed between transmission, reflection off a buried discontinuity, and reception back at a surface radar antenna. A pulse of radar energy is generated on a dipole transmitting antenna that is placed on, or near, the ground surface. The resulting wave of electromagnetic energy propagates downward into the ground, where portions of it are reflected back to the surface when it encounters buried discontinuities. The discontinuities where reflections occur are usually created by changes in electrical properties of the sediment or soil, variations in water content, lithologic changes, or changes in bulk density at stratigraphic interfaces. Reflection can also occur at interfaces between archaeological features and their surrounding soil or sediment. Void spaces in the ground, such as may be encountered in burials, tombs, tunnels, or caches, will also generate significant radar reflections due to changes in radar-wave velocity.

The depth to which radar energy can penetrate and the amount of definition that can be expected in the subsurface is partially controlled by the frequency of the radar energy transmitted. The radar-energy frequency controls both the wavelength of the propagating wave and the amount of weakening, or attenuation, of the waves in the ground. Standard GPR antennas propagate radar energy that varies in bandwidth from about 10 megahertz (MHz) to 1000 MHz. Antennas usually come in standard frequencies; each antenna has one center frequency, but it produces radar energy that ranges around that center by about two octaves. An octave is one half to two times the center frequency.

Radar antennas are usually housed in a fiberglass or wooden sled that is placed directly on the ground (Fig. 2), or supported on wheels a few centimeters above the ground. Antennas can also be placed directly on the ground without being housed in a sled. When two antennas are employed, one is used as a transmitting antenna and the other as a receiving antenna. When a single antenna is used as both a sender and receiver, it is turned on to transmit a radar pulse and then immediately switched to receiving mode in order to receive and measure the returning reflected energy.

Figure 2. Acquisition of GPR data using paired 300 megahertz (MHz) frequency antennas housed in a fiberglass sled. The reflection data are transmitted from the antenna to the control unit by the cable.

Antennas are usually hand-towed along survey lines within a grid at an average speed of about two kilometers per hour, or they can be pulled behind a vehicle at speeds of ten kilometers per hour or greater. There are a number of ways to move antennas across the ground while collecting data, depending on the type of GPR equipment being used and its accompanying software. In one method, the antennas move continuously along the ground while radar energy is constantly being transmitted into the ground at a set rate. Depending on the speed the antennas are moved, a greater or lesser amount of data may be collected per meter along a profile. Antennas may also be moved along a transect in steps, collecting data only at given surface locations equally spaced along a line. A greater coverage could be obtained by collecting data every five centimeters rather than every ten centimeters, but it would take twice as long to survey a line. Another data-acquisition method allows the operator to collect data at pre-programmed distances (every five centimeters, for example) along a transect while moving the antennas continuously. In this method, data is collected at equally spaced distances that are controlled by a distance-calibration wheel located on the antennas.

The authors of this book have worked primarily with GPR systems that send and receive data continuously, and all of the examples in the text were collected on those types of systems. The step method of data acquisition also has much to recommend it, including good subsurface resolution, image clarity, and excellent data for post-acquisition computer processing. Its acquisition method, however, necessitates more field time, and therefore less data can be acquired in a given amount of time. In the last few years, radar equipment manufacturers have been building their systems so that data can be collected by a number of methods, depending on the preference of the user or the characteristics a site.

The most efficient method in subsurface radar mapping is to establish a grid across a survey area prior to acquiring the data (Doolittle and Miller 1991). Usually rectangular grids are established with a line spacing of one meter or less. Rectangular grids produce data that are easier to process and interpret, but other types of grid acquisition patterns may be necessary because of surface topography or other obstructions. Surveys lines that radiate outward from one central area have sometimes been used, for instance to define a moat around a central fort-like structure (Bevan 1977). A rhomboid grid pattern was used with success within a sugarcane field on the side of a hill (Conyers 1995a), where antennas had to be pulled between planted rows. Data from nonrectangular surveys is just as useful as that acquired in rectangular grids, although more field time may be necessary in surveying and reflection data must be manipulated differently during computer processing and interpretation.

Occasionally GPR surveys have been carried out on the frozen surface of lakes or rivers (Annan and Davis 1977; Davis and Annan 1992; Doolittle and Asmussen 1992; Jol and Smith 1992). Radar waves will easily pass through ice and fresh water into the underlying sediment, revealing features on lake or river bottoms and in the subsurface. A radar sled can easily be floated across the surface of a lake or river and onto the shore, all the while collecting data from the sediment layers (Wright et al. 1984). These techniques, however, will not work in saltwater because the high electrical conductivity of the saline water will quickly dissipate the electromagnetic energy before it can be reflected back to the receiving antenna.

When the antennas are pulled continuously along a transect line within a pre-surveyed grid, continuous pulses of radar energy are sent into the ground and reflected off subsurface discontinuities; the reflected pulses are then received and recorded back at the surface. The movable radar antennas are connected to the control unit by cable. Some systems record the reflection data digitally at the antenna, and the digital signal is then sent through fiber-optic cables back to the control module (Davis and Annan 1992). Other systems send an analog signal from the antennas, through coaxial copper cables, to the control unit, where it is then digitized. Older GPR systems, which lack the capability to digitize the reflection signals in the field, must record reflection data on magnetic tape or paper records.

The reflection data—the two-way travel time and the amplitude and wavelength of the reflected radar waves derived from the pulses—are amplified, processed, and recorded for immediate viewing or later post-acquisition processing and display. During field-data acquisition, the radar transmission process is repeated many times a second as the antennas are pulled along the ground surface or moved in steps. When the composite of all reflected wave traces along the transect is displayed, a cross-sectional view of significant subsurface reflection surfaces is generated (Fig. 1). In this fashion, two-dimensional profiles, which approximate vertical "slices" through the earth, are created along each grid line.

Radar reflections are always recorded in "two-way time," which is the time it takes a radar wave to travel from the surface antenna into the ground, be reflected off a discontinuity, travel back to the surface, and be recorded. In archaeology, one of the advantages of GPR surveys over other methods is that the subsurface stratigraphy and archaeological features at a site can be mapped in real depth. This is possible because the two-way travel time of radar pulses can be converted to depth if the velocity of the radar wave's travel through the ground is known.

The propagation velocity of radar waves projected through the earth depends on a number of factors, the most important one being the electrical properties of the materials through which they pass (Olhoeft 1981). In air, radar waves travel at the speed of light, which is approximately thirty centimeters per nanosecond (one nanosecond is one billionth of a second). When radar energy travels through dry sand, its velocity slows to about fifteen centimeters per nanosecond. If the radar energy were then to pass through a water-saturated sand unit, its velocity would slow further to about five centimeters per nanosecond or less. At each of these surface interfaces where velocity changes, reflections are generated.

GENERATION, PROPAGATION, AND REFLECTION OF RADAR ENERGY

The primary goal of most GPR investigations in archaeology is to differentiate subsurface interfaces. All sedimentary layers in the earth have particular electrical and magnetic properties that affect the rate of electromagnetic energy propagation and its dissipation in the ground. The greater the contrast of these properties between two buried materials, the stronger the reflected signal (Sellman et al. 1983). The difficulty of measuring the electrical and magnetic properties of buried units where there are few, if any, excavations, usually precludes accurate calculations of reflectivity, and therefore only estimates can usually be made.

The strongest radar reflections in the ground usually occur at the interface of two thick layers with greatly varying electrical properties. The degree to which radar reflections can be "seen" on profiles is related to the amplitude of the reflected waves. The higher the amplitude, the more visible the reflections. Lower-amplitude reflections usually occur when there are only small differences in the electrical properties between layers.

Two-dimensional computer-generated models, discussed in chapter 5, can be used as a guide to understanding the reflection amplitudes and the resulting resolution of various buried features. If enough information is available about potential targets, stratigraphy, and soil conditions, these models can predict the intensity of reflections and their placement in space. Synthetic modeling programs are one of the newest advances in GPR technology with wide archaeologic applications. They not only can predict whether archaeological features will generate visible reflections, but also be manipulated for varying geologic conditions and radar equipment.

Reflected radar waves that are received at the surface antenna are converted to electrical signals, which are manifested as minor changes in voltage. These signals are transmitted to the control unit, amplified, and recorded. One of the benefits of digitally recorded data is that they can be processed, filtered, and viewed immediately on a computer screen in the field; they are also more readily processed when transferred to a personal computer once back in the office. Little can be done in the way of enhancement when raw reflection data are only recorded as paper records (Milligan and Atkin 1993).

Radar energy becomes both dispersed and attenuated as it radiates into the ground. When portions of the original transmitted signal are reflected back toward the surface, they will suffer additional attenuation by the material through which they pass before finally being recorded at the surface. Therefore, to be detected as reflections, important subsurface interfaces not only must have sufficient electrical contrast at their boundary, but also must be located at a shallow enough depth that sufficient radar energy is still available for reflection back to the surface. As radar energy is propagated to increasing depths and the signal becomes weaker and is spread out over more surface area, less energy is available for reflection, and it is possible that only very low amplitude waves will be recorded. For every site, the maximum depth of resolution will vary with the geologic conditions and the equipment being used. Data filtering and other data-amplification techniques that will enhance very low amplitude reflections generated deeply in the ground can sometimes be applied to reflection data after acquisition in order to make them more visible.

PRODUCTION OF CONTINUOUS REFLECTION PROFILES

Most radar units used for archaeological investigation transmit short discreet pulses into the earth and then measure the reflected waves derived from those pulses as the antennas are moved along the ground. A series of reflected waves is recorded as the antennas are moved along a transect. If a greater degree of subsurface resolution is necessary, the radar antennas can be pulled more slowly and more reflection traces can be recorded for every centimeter of ground crossed. If less subsurface coverage is necessary, the antennas can be moved faster. If the step method of acquisition us used, the distance between steps can be lengthened or shortened, depending on the subsurface resolution desired. The radar control unit can also be programmed to produce a greater or smaller number of radar pulses per distance moved.

Many of the older radar units, which recorded data continuously along a transect, were set up to record at standard rates—12.8, 25.6, 51.2, or 102.4 reflected traces per second—that were adjusted with a manual switch on the control unit. (One "trace" is a complete reflected wave that is recorded from the surface to whatever depth is being surveyed.) The computer software in modern GPR units allows the recording rate to be adjusted to any rate depending on the precision necessary. For instance, if one reflection trace is desired every two centimeters along a transect and the anticipated speed of the antennas along the ground surface is twenty centimeters per second, then the recording rate could be adjusted to record ten traces per second. If one reflection trace is desired every one centimeter, then the recording rate could be doubled to twenty traces per second.

As reflections from the subsurface are recorded in distinct traces and plotted together in a profile, a two-dimensional representation of the subsurface can be made (Fig. 1). A series of reflections that together create a horizontal or sub-horizontal line on a profile (either dark or light in standard black and white or grey-scale profiles) is usually referred to as "a reflection." A distinct reflection visible in profiles is usually generated from a subsurface boundary such as a stratigraphic layer or from some other physical disconti- nuity such as a water table. Reflections recorded later in time are usually those received from deeper in the ground. There can also be "point source" reflections that are generated from one feature in the subsurface. These are visible on two-dimensional profiles as hyperbolas. Due to the wide angle of the transmitted radar beam, the antenna will "see" the point source prior to arriving directly over it, and continue to "see" it after it has passed (Fig. 3). The resulting recorded reflection will therefore create a reflection hyperbola, sometimes incorrectly called a diffraction, on two-dimensional profiles.

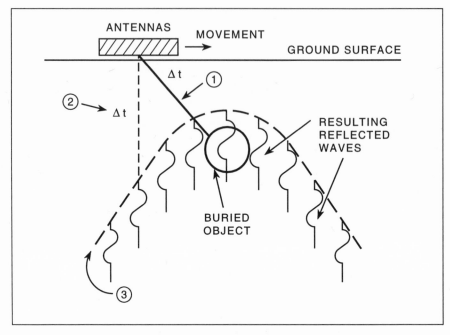

Figure 3. Generation of a reflection hyperbola from a "point source." As the antenna is moved along the ground surface, its wide field of view allows the antenna to "see" the point source prior to reaching it (1). The reflection time (Δt), however, is recorded as if the point source were directly beneath the antenna (2). Only when the antenna is directly over the object will it record the correct time. As it moves away from the point source, reflections will continue to be recorded from the same point, creating a two-dimensional hyperbolic reflection in the resulting profile (3).

DATA RECORDING

Some of the earlier-model GPR systems were only able to record reflection data on paper by means of a graphic recorder (Batey 1987; Fischer et al. 1980; Loker 1983), which uses electro-sensitive paper that moves across an electrically-charged moving stylus. The stylus is fed the amplified incoming reflection data from the subsurface. When the reflected wave amplitudes reach a preselected threshold, the surface of the paper is burned off, leaving dark and light lines. The electrical energy transmitted to the stylus varies with amplitude changes of the reflected waves and darker and lighter variations will occur for each reflection. A depth scale, in two-way travel time measured in nanoseconds, is usually imposed on the paper printouts. When using a graphic recorder, the operator can vary the speed of the paper and the moving stylus to produce a wide variety of profile styles (Batey 1987; Fischer et al. 1980). This type of data recording has nearly been superseded by digital systems, but a few of these systems are still in operation.

In some of the more antiquated analog GPR units, reflected wave data can be recorded on magnetic tape as small voltage changes around an arbitrary mean (Loker 1983). This kind of recorded data can be reprocessed to convert the analog signal to digital data using a computer digitizer.

In the early 1980s, GPR units were invented that recorded GPR reflections internally as digital data (Annan and Davis 1992; Geophysical Survey Systems, Inc. 1987). In these units, a computer built into the control unit allows the resulting data to be edited, processed, filtered, and corrected easily. Digital units have become the standard equipment in most GPR surveys, although good data can still be acquired with the older analog units.

PHYSICAL PARAMETERS AFFECTING RADAR TRANSMISSION

The maximum effective depth of penetration of GPR waves is a function of the frequency of the waves that are propagated into the ground and the physical characteristics of the material through which they are traveling. The physical properties that affect the radar waves as they pass through a medium are the electrical conductivity, and the magnetic permeability (Annan et al. 1975; Geophysical Survey Systems, Inc. 1987). Soils, sediment, or rocks that are "dielectric" will permit the passage of most electromagnetic energy without actually dissipating it. The more electrically conductive a material is, the less dielectric it is. For maximum radar-energy penetration, a medium should be highly dielectric with a low electrical conductivity.

Relative dielectric permittivity (RDP) of a material is defined as the capacity of a material to store, and then allow the passage of, electromagnetic energy when a field is imposed upon it (von Hippel 1954). It can also be thought of as a measure of the ability of a material within an electromagnetic field to become polarized, and therefore respond to, propagated electromagnetic waves (Olhoeft 1981). This measurement is the standard unit used to measure radar propagation in the ground. RDP is calculated as the ratio of a material's electrical permittivity to the electrical permittivity of a vacuum (which is one). Dielectric permittivities of materials vary with their composition, moisture content, bulk density, porosity, physical structure, and temperature (Olhoeft 1981). It is usually difficult to calculate RDP in the field, but it can be estimated using a number of techniques discussed in chapter 6. It can also be measured in the laboratory from soil and sediment samples.

The relative dielectric permittivity in air, which exhibits only negligible electromagnetic polarization, is approximately 1.0003 (Dobrin 1976), and is usually rounded to one. In volcanic or other hard rocks, it can range from 6 to 16; and in wet soils or clay-rich units, it can approach 40 or 50. In unsaturated sediment with little or no clay, relative dielectric permittivities can be 5 or lower. In general, the higher the RDP of a material, the slower the velocity of radar waves passing through it. An estimation of some RDPs for some soil, sediment, and rock types is shown in table 1.

If data are not immediately available about field conditions, the RDP can only be estimated. Methods described in chapter 6 allow an accurate measurement of RDP in the laboratory, using specialized equipment, or in the field, using a number of different velocity analysis techniques. Equation 1, which relates RDP and the velocity of the radar wave as it passes through a material, is shown below. Practitioners of GPR often casually use phrases like "the velocity of a material" that, to the uninitiated, make it appear that the material is moving.

Table 1

Typical Relative Dielectric Permittivities (RDP) of Common Geological Materials, Using a 100 MHz Antenna

Material	RDP	Material	RDP
Air	1	Dry Silt	3–30
Freshwater	80	Saturated Silt	10–40
Ice	3–4	Clay	5–40
Seawater	81–88	Permafrost	4–5
Dry Sand	3–5	Average Surface Soil	12
Saturated Sand	20–30	Dry, Sandy	
Volcanic ash/pumice	4–7	Coastal Land	10
Limestone	4–8	Forested Land	12
Shale	5–15	Rich Agricultural Land	15
Granite	5–15	Concrete	6
Coal	4–5	Asphalt	3–5

Source: Modified from Davis and Annan (1989) and Geophysical Survey Systems, Inc. (1987).

What is actually meant by this phrase is the velocity of the radar waves *through* the material.

$$\sqrt{K} = \frac{C}{V}$$

K = relative dielectric permittivity (RDP) of the material through which the radar energy passes
C = speed of light (.2998 meters per nanosecond)
V = velocity of the radar energy as it passes through a material (measured in meters per nanosecond)

Equation 1. Relative Dielectric Permittivity and Radar Velocity Relationship in GPR.

The greater the difference between the relative dielectric permittivity of materials in the subsurface, the larger the amplitude of the reflection generated. The magnitude of the reflection generated at an interface can be quantified using equation 2 if the RDP of the two materials is known (Sellman et al. 1983; Walden and Hosken 1985).

$$R = [\sqrt{(K_1)} - \sqrt{(K_2)}] \ [\sqrt{(K_1)} + \sqrt{(K_2)}]$$

R = coefficient of reflectivity at a buried surface
K_1 = RDP of the overlying material
K_2 = RDP of the underlying material

Equation 2. The Coefficient of Reflectivity at an Interface between Materials of Differing Relative Dielectric Permittivity.

In order to generate a significant reflection, the change in dielectric permittivity between two materials must occur over a short distance. When the RDP changes gradually with depth, only small differences in reflectivity will occur every few centimeters, and only weak reflections—or no reflection at all—will be generated.

Two other physical parameters affect radar-wave transmission through the ground are the magnetic permeability of the medium and its electrical conductivity. Magnetic permeability is a measure of the ability of a medium to become magnetized when an electromagnetic field is imposed upon it (Sheriff 1984). Most soils and sediments are only slightly magnetic, and therefore have a low magnetic permeability. The higher the magnetic permeability, the more electromagnetic energy will be attenuated during its transmission. Media that contain magnetite minerals, iron-oxide cement, or iron-rich soils can have a high magnetic permeability and therefore transmit radar energy poorly.

Electrical conductivity is the ability of a medium to conduct an electrical current (Sheriff 1984). When a medium through which radar waves are passing has a high conductivity, radar energy will be highly attenuated. In a highly conductive medium, the electrical component of the electromagnetic energy is essentially being conducted away into the earth and becomes lost. This occurs because the electrical and magnetic fields are constantly "feeding" on each other during transmission. If one is lost, the total field dissipates. Highly conductive media include those that contain saltwater and those with a high clay content, especially if the clay is wet. Any soil or sediment that contains soluble salts or electrolytes in the groundwater will also have a high electrical conductivity. Agricultural runoff that is partially saturated with soluble nitrogen and potassium can raise the conductivity of a medium, as will wet calcium carbonate–impregnated soils or caliche soils in desert regions.

Radar energy will not penetrate metal. A metal object will reflect one hundred percent of the radar energy that strikes it and will shadow anything directly beneath it.

RADAR PROPAGATION

Many ground-penetrating radar novices envision the propagating radar pattern as a narrow pencil-shaped beam that is focused directly down from the antenna. In fact, GPR waves produced by standard commercial antennas radiate radar energy into the ground in an elliptical cone (Fig. 4) with the apex of the cone at the center of the transmitting antenna (Annan and Cosway 1992; Annan and Cosway 1994; Arcone 1995; Davis and Annan 1989). This elliptical cone of transmission is usually elongated parallel to the direction of antenna movement along the ground surface. The radiation pattern is generated from a horizontal electric dipole antenna to which elements called shields are sometimes added that effectively reduce upward radiation. Sometimes the only shielding mechanism is a metal plate that is placed above the antenna to re-reflect radiating energy upward. Because of cost and portability (size and weight) considerations, the use of more complex radar antennas that might be able to focus energy into the ground more efficiently and in a narrower beam has been limited.

When an electric dipole antenna is located in air (or supported within the antenna housing), the radiation pattern is approximately perpendicular to the long axis of the antenna. When this dipole antenna is placed on the ground, a major change in the radiation pattern occurs due to ground coupling (Engheta et al. 1982). Ground coupling is the ability of the electromagnetic field to move from transmission in the air into the ground. During this process, refraction, which occurs as the radar energy passes through surface units, causes a change in the directionality of the radar beam, with most of the energy channeled downward in a cone from the propagating antenna (Annan et al. 1975). The higher the RDP of the surface material, the lower the velocity of the transmitted radar energy, and the more focused (less broad) the conical transmission pattern becomes as it moves into the ground (Goodman 1994). This focusing effect continues to occur as radar waves travel into the ground and material of higher and higher RDP is encountered (Fig. 5). The amount of energy refraction that occurs with depth, and therefore the amount of focusing, is a function of Snell's Law (Sheriff 1984). In Snell's Law, the amount of reflection or re-

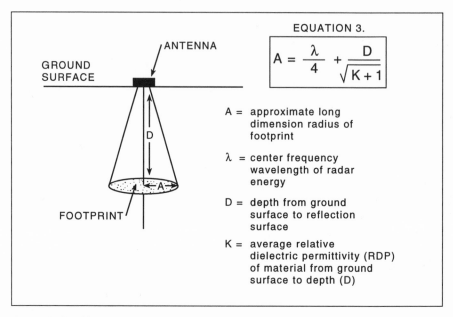

Figure 4. The elliptical cone of GPR penetration into the ground. Equation 3 defines the cone geometry and the "footprint" radius at different depths. The footprint is the area illuminated on a buried horizontal surface. (Modified from Annan and Cosway 1992.)

fraction that will occur at a boundary between two media depends on the angle of incidence and the velocity of the incoming waves. In general, the greater the increase in RDP with depth, the more focused the cone of transmission becomes. The opposite can also occur if materials of gradually lower RDP are encountered as radar waves travel into the ground (Fig. 5). In this case, the cone of transmission would gradually expand outward as refraction occurs at each interface.

Radiation fore and aft from the antenna is usually greater than to the sides, making the "illumination" pattern on a horizontal subsurface plane approximately elliptical in shape (Fig. 4), with the long axis of the ellipse parallel to the direction of antenna travel (Annan and Cosway 1992). In this way, the subsurface radiation pattern on a buried horizontal plane (sometimes called the "footprint") is always "looking" not only directly below the antenna but also in front, behind, and to the sides as it travels across the ground.

In order to minimize the amount of reflection data derived from the sides of a survey line, the long axes of the antennas are aligned perpendicular to the survey line. This allows the cone of transmission to be elongated in an in-line direction (Fig. 4). If there are narrow elongated features in the subsurface that are parallel to the direction of antenna travel (and therefore parallel to the electrical field generated by the antenna), only a small portion of the radar energy will be reflected back at the surface. Elongated buried features of this sort would usually have to be oriented perpendicular to the direction of travel in order to be visible on GPR profiles. There are various other antenna orientations that will achieve different subsurface search patterns, but most of these are not used in standard GPR surveys for archaeology (Annan and Cosway 1992).

Some antennas, especially those in the low-frequency range from 80 to 120 MHz, are not shielded and will therefore radiate radar energy in all directions. Unshielded antennas can generate reflections from a person pulling the radar equipment or from any other objects nearby, such as trees, buildings, metal fences, or power lines (Lanz et al. 1994). Discrimination of individual targets, especially those of interest in the subsurface, can be difficult if these types of antennas are used. However, if the unwanted reflections generated from unshielded antennas all occur at approximately the same time, such as those off a person pulling the antennas, they can easily be filtered out later, assuming the data are recorded digitally. An example of this type of data filtering from an unshielded antenna is included in chapter 4. If reflections are recorded from randomly located trees, surface obstructions, or people

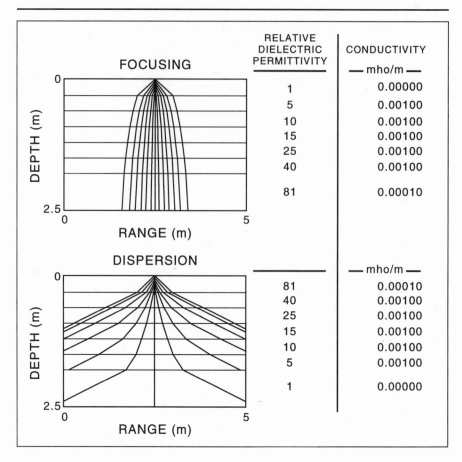

Figure 5. The focusing effect (upper figure) of radar waves traveling through layers of increasing relative dielectric permittivity (RDP) and therefore decreasing velocity. If radar waves were to travel through layers of increasing velocity, and therefore decreasing RDP (lower figure), the waves would tend to disperse with depth.

moving about near the antenna, they usually cannot be discriminated easily from important subsurface reflections, and interpretation of the data is much more difficult.

If the radiating antenna is shielded so that energy is being propagated in a mostly downward direction, the angle of the conical radiation pattern can be estimated, depending on the center frequency of the antenna being used (Annan and Cosway 1992). An estimation of this radiation pattern is especially important when designing line spacing within a grid so that all subsurface features of importance are "illuminated" by the transmitted radar energy and can therefore generate reflections. In general, the angle of the cone is defined by the relative dielectric permittivity of the material through which the waves pass and the frequency of the radar energy emitted from the antenna.

An equation that can be used to estimate the width of the transmission beam at varying depths (the footprint) is shown in figure 4. This equation (Equation 3) can only be used as a rough approximation of real-world conditions because it assumes a consistent dielectric permittivity of the medium through which the energy passes. Outside of strictly controlled laboratory conditions, this is almost never the case. Sedimentary and soil layers within the earth have variable chemical constituents and differences in retained moisture, compaction, and porosity. These and other variables can create a complex layered system with varying dielectric permittivities and therefore differing energy transmission patterns. Transmission beam dimensions can also be affected by antenna design differences.

Any estimation of the orientation of transmitted energy is also complicated by the knowledge that radar energy propagated from a surface antenna is not one distinct frequency but can range many hundreds of megahertz around a center frequency. If one were to make a series of calculations on each layer, assuming all the variables could be determined and also assuming one distinct antenna frequency, then the "cone" of transmission would be seen to widen in some layers and narrow in others, creating a very complex three-dimensional pattern. The best one can usually do for most archaeological applications is to make an estimation of the radar-beam configuration based on approximate field conditions. Some determination of the propagation-beam dimensions is important prior to conducting a survey so that grid lines can be spaced at distances smaller than the maximum cone width at the depth necessary to delineate the features of interest (Equation 3 in Fig. 4). Any wider spacing of survey lines may allow important subsurface features to go undetected.

Figure 6 is a graph of differing footprint radii for a 300 MHz center-frequency antenna at varying RDPs. As can be seen from this graph, the footprint dimension gets larger with depth as the cone of transmission spreads out. The footprint is much larger when radar energy travels through a material with a low RDP. Higher-RDP material tends to focus the beam of transmission, decreasing the radius of the subsurface footprint. Therefore, when conducting a

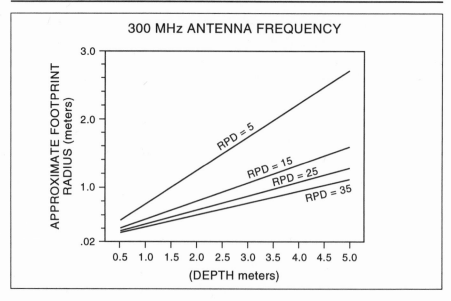

Figure 6. Approximate radius of "footprints" at varying depths and relative dielectric permittivities of material for a 300 MHz antenna. The higher the RDP of the material, the more focused the radar beam becomes in the ground, and therefore the smaller the footprint radius.

survey in areas with a high RDP, transect lines should be more closely spaced in order to make sure all subsurface features are illuminated with radar energy.

ANTENNA FREQUENCY CONSTRAINTS

One of the most important decisions in ground-penetrating radar surveys is the selection of antennas with the correct operating frequency necessary for the depth and the resolution of the features of interest (Huggenberger et al. 1994; Smith and Jol 1995). Commercial GPR antennas range from about 10 to 1000 megahertz (MHz) center frequency (Annan and Cosway 1994; Fenner 1992; Malagodi et al. 1996; Olson and Doolittle 1985). General purpose GPR systems use dipolar antennas that typically have a two octave bandwidth, meaning that the frequencies vary between one-half and two times the dominant frequency. For example, a 300 MHz center-frequency antenna generates radar energy with wavelengths ranging from about 150 to 600 MHz. The frequency distribution of an idealized 500 MHz wave, which is a reasonable approximation of a GPR pulse, is shown in figure 7. The frequency distribution is not a bell-shaped curve around a mean, but an asymmetrical distribution around a dominant frequency, in this case not actually 500 but 516 MHz.

Figure 8 shows the actual frequency distribution derived from a radar pulse created from a 500 MHz antenna. In this antenna, the transmitted pulse that was recorded is somewhat "noisy," with the beginning of one strong pulse, recorded at two nanoseconds, followed by antenna "ringing" and system noise after about six nanoseconds. This actual antenna test is anything but the clean, idealized pulse shown in figure 7. The frequency distribution of this test varies between 200 and about 800 MHz, with many dominant frequencies recorded as "spikes" in the frequency distribution. These variations in dominant frequencies may be caused by irregularities in the antenna's surface or other electronic components located within the antenna system. These types of variations are common in all antennas, and each has its own irregularities that produce different pulse signatures and different dominant frequencies. It must be remembered that just because an antenna is identified as one frequency does not necessarily mean that it will produce radar energy with a center at exactly that frequency. If there is any question as to frequency, a frequency distribution test that will yield a distribution curve like that shown in figure 8 should be performed prior to collecting data.

This already confusing transmission-frequency situation is further complicated when radar energy is propagated into the ground. When radar waves move through the ground, the center frequency typically "loads down" to a lower dominant frequency (Engheta et al. 1982). The new propagation frequency will vary depending on the electrical properties of near-surface soils and sediment that change the velocity of propagation and the amount of "coupling" of the propagating energy with the ground. At present there is little hard data that can be used to predict accurately what the "downloaded" frequency of any antenna will be under varying conditions. For most archaeological purposes, it is only important to be aware that there is a downloading effect that can change the dominant radar frequency and affect calculations of subsurface transmission patterns, penetration depth, resolution, and other parameters.

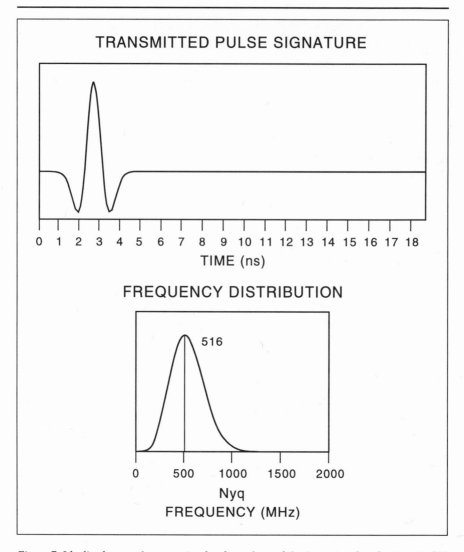

Figure 7. Idealized wave of a transmitted radar pulse and the frequency distribution of a 500 MHz antenna. The upper figure is the idealized pulse as it is transmitted into the ground over time. The lower figure is the calculated frequency distribution of that pulse, with a calculated center frequency of 516 MHz. (Modified from Powers and Olhoeft 1995.)

Proper antenna center-frequency selection can, in most cases, make the difference between success and failure in a GPR survey, and it must be planned for in advance. In general, the greater the necessary depth of investi-

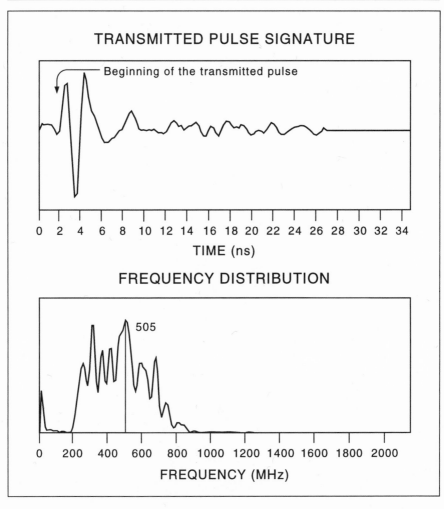

Figure 8. Actual 500 MHz transmitted pulse and its frequency distribution. The wide bandwidth of the antenna generates frequencies from 200 to 850 MHz, with many peaks, and a center frequency of 505 MHz. (Courtesy of Mike Powers.)

gation, the lower the antenna frequency that should be used. Lower-frequency antennas are much larger, heavier, and more difficult to transport to and within the field than high-frequency antennas. One model of an 80 MHz antenna used for continuous GPR acquisition is larger than a fifty-five-gallon oil drum cut in half lengthwise, and it weighs between 125 and 150 pounds (Fig. 9). It

Figure 9. GPR antennas of different frequencies. From the bottom up, the frequencies are 900, 500, 300, 100, and 80 MHz.

is not only difficult to transport to and from the field, but usually must be moved along transect lines using some form of wheeled vehicle or sled. In contrast, a 900 MHz antenna is smaller than a shoe box, weighs very little, and can easily fit into a suitcase (Fig. 9). Some lower-frequency antennas used for acquiring data in the step method are not nearly as heavy as those used in continuous data acquisition, but they are equally as unwieldy.

Low-frequency antennas (10–120 MHz) generate long-wavelength radar energy that can penetrate up to fifty meters in certain conditions, but they are capable of resolving only very large subsurface features. In pure ice, antennas of this frequency have been known to transmit radar energy many kilometers. In contrast, the maximum depth of penetration of a 900 MHz antenna is about one meter or less in typical soils, but its generated reflections can resolve features down to a few centimeters. A trade-off therefore exists between depth of penetration and subsurface resolution. The dominant wavelength for different center-frequency antennas and the changes in wavelength as it passes through materials with differing RDPs is shown in table 2.

Table 2

The Center-Frequency Wavelengths of Differing Radar Antennas and the Propagated Wavelength Changes in Materials with Differing Relative Dielectric Permittivities (RDP)

Antenna Center Frequency (MHz)	Center Frequency Wavelength in air (meters)	Center Frequency Wavelength in a Medium with		
		RDP = 5	RDP = 15	RDP = 25
1000	0.30	0.13	0.08	0.06
900	0.33	0.15	0.09	0.07
500	0.60	0.27	0.15	0.12
300	1.00	0.45	0.26	0.20
120	2.50	1.12	0.65	0.50
100	3.00	1.34	0.77	0.60
80	3.75	1.68	0.97	0.75
40	7.50	3.35	1.94	1.50
32	9.38	4.19	2.42	1.88
20	15.00	6.71	3.87	3.00
10	30.00	13.42	7.75	6.00

The depth of penetration and the subsurface resolution are actually highly variable and depend on many site-specific factors such as overburden composition, porosity, and the amount of retained moisture, so the values shown in table 2 must be taken only as a rule of thumb.

Some of the important factors that must be considered in choosing an antenna frequency are:

- Electrical and magnetic properties of the host environment of the site.
- Depth of study necessary.
- Size and dimensions of the archaeological features that must be resolved.
- Site access.
- Presence of possible external electrical interference of the same wavelength as the radar waves that will be propagated into the ground.

At higher frequencies, usually greater than 1500 MHz, some geologic materials containing water will exhibit higher signal attenuation due to energy loss from molecular relaxation (Annan and Cosway 1994; Olhoeft 1994a). This is usually not a condition that affects most GPR surveys because the most commonly used antenna frequencies are lower than this.

If large amounts of clay, and especially wet clay, are present, then attenuation of the radar energy with depth will occur very rapidly (Doolittle and Miller 1992; Duke 1990; Keller 1988; Shih and Doolittle 1984). Attenuation can also occur if sediment or soils are saturated with salty water, especially seawater (van Heteren et al. 1994).

The ability to resolve buried features is mostly determined by frequency and therefore by the wavelengths of the radar energy being transmitted into the ground. The wavelength necessary for resolution varies depending on whether a three-dimensional object or an undulating surface is being investigated. For GPR to resolve three-dimensional objects, reflections from at least two surfaces, usually a top and bottom interface, need to be distinct. Resolution of a single buried planar surface, however, needs only one distinct reflection, and therefore wavelength is not as important.

An 80 MHz antenna generates an electromagnetic wave of about 3.75 meters in length when transmitted in air (Table 2). When the wavelength in air is divided by the square root of the RDP of the material through which it passes, an estimate of the subsurface wavelength can be made. For example, when an 80 MHz wave travels through material with a RDP of 5, its wavelength decreases to about 1.7 meters (Table 2). The 300 MHz antenna generates a radar wave with a wavelength of one meter in air, which decreases to about 45 centimeters in material with a RDP of 5. In order to distinguish reflections from two parallel planes (the top and bottom of a buried object, for instance) they must be separated by at least one wavelength of the radar energy that is passing through the ground (Davis and Annan 1989). If the two reflections are not separated by one wavelength, then the resulting reflected waves from the top and bottom will either be destroyed or unrecognizable due to constructive and destructive interference, as illustrated in figure 10. When two interfaces are separated by greater than one wavelength, however, two distinct reflections are generated, and the feature can be resolved.

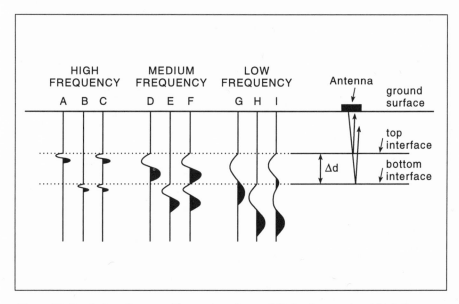

Figure 10. Resolution of a top and bottom interface at differing frequencies. A high-frequency wave defines the top (A) and bottom interface (B). The resulting wave from these reflections (C) can resolve both interfaces because the distance between the two (Δd) is greater than the wavelength. A medium-frequency wave (D) defines the top interface and the bottom interface (E), and the resulting wave (F) can just barely define both interfaces because the wavelength approaches Δd. A low-frequency wavelength defines the top interface (G), but because the bottom interface is separated from the top interface by less than the wavelength (H), the resulting wave (I) is created by interference between the two waves and only the top interface can be resolved.

If only one buried planar surface is being mapped and it generates a significant reflection that cannot be confused with others, then the first reflected waves produced from that layer will be visible and can be measured accurately, independent of the radar wavelength. This can be more difficult when the surface is highly irregular or undulating. Subsurface reflections of buried surfaces generated by longer-wavelength radar waves tend to be less sharp when viewed in a standard GPR profile, and therefore many small irregularities on the buried surface are not visible. This occurs because the conical radiation pattern of an 80 MHz antenna is about three times broader than that of the 300 MHz antenna (Annan and Cosway 1992). The reflected data received back at the surface from the lower-frequency antenna has therefore been reflected from a much greater subsurface area, which has the result of averaging out the low percentage of reflections that occurred from the smaller irregular features. A reflection profile produced from reflections by an 80 MHz antenna will therefore produce an average, but less accurate, representation of a buried surface. In contrast, a 300 MHz transmission cone is about three times narrower than an 80 MHz radar beam, and its resolution of subsurface features on the same buried surface is much greater. This phenomenon is demonstrated dramatically in figure 11, where one buried surface is visible in both 80 and 300 MHz center-frequency data along approximately the same traverse. Although the buried surface is visible in both profiles, the ability of GPR to delineate subtle irregularities is greatly improved with the higher-frequency data.

Radar energy that is reflected off a buried subsurface interface that slopes away from a surface transmitting antenna will probably not travel back to the receiving antenna. In this case, all the reflected energy will be lost, and the sloping interface will go unnoticed in reflection profiles. A buried surface with this orientation will only be visible if an additional traverse is located in an orientation where it is sloping toward the surface antennas. This is one reason why it is always important to acquire lines of data within a closely spaced surface grid.

The amount of reflection that will occur off a buried feature is also determined by the ratio of the object dimension to the wavelength of the radar wave in the ground. Short-wavelength (high-frequency) radar waves are capable of resolving very small features, but they will not penetrate to a great depth. Longer-wavelength radar energy will resolve only larger features, but it will penetrate deeper in the ground.

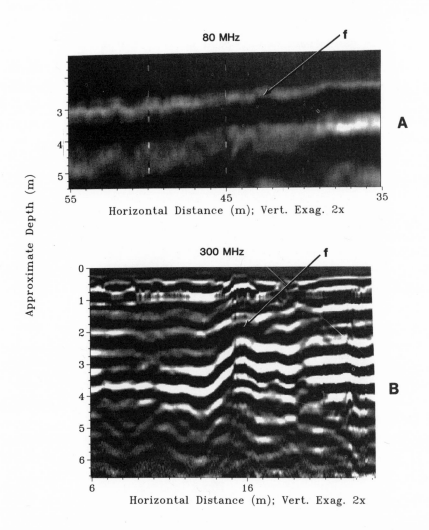

Figure 11. Subsurface resolution of a buried planar interface at two different frequencies. (A): The subsurface topographic feature (f) is visible as a minor change in slope. It can barely be resolved in the 80 MHz profile. (B): The same feature (f) on a parallel 300 MHz line illustrates the increase in resolution with a higher frequency antenna. These profiles imaged the buried TBJ living surface at the Ceren Site, El Salvador.

Some features in the subsurface may be described as "point targets," while other are more similar to planar surfaces. Planar surfaces can be stratigraphic, like soil horizons, or large flat archaeological features, such as pithouse floors. Point targets are features such as tunnels, voids, artifact caches, or any other non-planar object. Depending on a planar surface's thickness, reflectivity, orientation, and depth of burial, it is potentially visible with any frequency data, constrained only by the conditions discussed above. Point sources, however, can have little surface area to reflect radar energy and therefore are usually difficult to identify and map. They are sometimes indistinguishable from the surrounding material, but many times can be visible as small reflection hyperbolas on one line within a grid. Point targets can sometimes be visible using high-frequency antennas, as long as they are not buried too deeply.

In most geological and archaeological settings, the materials through which radar waves pass may contain many small discontinuities that reflect energy. These can only be described as clutter (that is, if they are not the target of the survey). The ability to image clutter is totally dependent on the wavelength of the radar energy being propagated. If both the features to be resolved and the discontinuities producing the clutter are on the order of one wavelength in size, then the reflection profiles will appear to contain only clutter and no discrimination can be made between the two. Clutter can also be produced by large discontinuities, such as cobbles and boulders, but only when a lower-frequency antenna producing a long wavelength is used. In all cases, the dimensions of features to be resolved and their surrounding matrix will determine the frequency of antenna to be used in a survey. Most often, if the target feature is large and not an extensive planar surface, the wavelength of the transmitted radar energy should be greater than the maximum dimension of the potential surrounding clutter in order for that feature to be resolved.

Buried features, whether planar or point sources, also cannot be too small in relation to their depth of burial because they will be undetectable. As a basic guideline, the cross-sectional area of the target to be illuminated should approximate the size of the footprint at the target depth (Equation 3 in Fig. 4). If the target is much smaller than the footprint size, then only a fraction of the reflected energy that is returned to the surface will have been reflected off the buried feature (Fig. 5) and any reflections returned from the buried feature may be indistinguishable from background reflections, and invisible on reflection profiles. Small features of this sort may still be discovered, however, but only after the raw data are processed using the amplitude slice mapping techniques discussed in chapter 8.

In order to determine whether a specific antenna frequency will be capable of resolving features of a known size at a known depth, calculations need to be done in advance using equation 3. Using this equation, an estimate of the footprint size at the desired depth can be calculated, but only if the RDP of the material can be estimated. If no accurate measurements are available for the RDP of the overburden, then estimates must be made from the values supplied in table 1.

The selection of antenna frequency is also dependent to some extent on site-area logistics. Dipole antennas vary greatly in size, depending on their center frequency. A 10 MHz antenna is about 15 meters long, a 100 MHz antenna is 1.5 meters, and a 1000 MHz is only 15 centimeters. As a result of these size differences, it is easy to see that the 1000 MHz center-frequency antenna can be easily moved around in almost any space, while the 10 MHz requires a large open area. It is important to select an antenna frequency with the highest possible frequency to attain the greatest resolution, taking into account the depth of penetration necessary, the size of the features to be resolved, and the ability to transport the antenna to, and within, the field. A compromise sometimes must be made between what is ideal for the site conditions and what is practical. A consideration of all these variables may preclude the use of GPR at some sites, which is always good to know before expending time and money preparing to conduct a GPR survey.

Ground-penetrating radar employs electromagnetic energy at frequencies that are similar to those used in television, FM radio, and other radio communication bands. If there is an active radio transmitter in the vicinity of the survey, then there may be some interference with the recorded signal. Most radio transmitters, however, have a quite narrow bandwidth and, if known, it can be determined in advance and an antenna frequency can be selected that is as different as possible from any frequency that might generate spurious signals in the reflected data. With the wide bandwidth of most GPR systems, it is usually difficult to avoid such external transmitter effects completely, and any major adjustments in antenna frequency may affect the survey objectives. Usually this only becomes a problem if the site is located near a military base or airport, or near radio transmission antennas. Walkie-talkies in use nearby during the acquisition of GPR data can also create FM noise in recorded reflection data; these should not be used during data collection. This type of radio noise can also be filtered out during post-acquisition data processing.

In summary, the following steps must be considered prior to selecting an antenna frequency:

- Obtain as much information as possible about the electrical and magnetic properties of a site. If this cannot be determined from direct field measurements, the type of soil and geologic materials should be known in advance so estimates of RDP can be made.
- Define the depth of resolution that is necessary to define the archaeological features of interest. Using estimates of RDP, the cone of transmission (Equation 3 in Fig. 4) can be predicted and potential resolution can be estimated from the footprint size at different frequencies.
- Decide whether or not it is physically possible to use the selected antenna frequency at the site to be surveyed. Transportability to and from the site and deployment over and around obstacles and obstructions once surveying is begun must be accounted for.
- If it is known that there is a substantial amount of radio interference present at a site, and if the source can be identified, then it may be appropriate to choose an alternate antenna frequency so as to minimize that influence. In general, this is not a simple task because it is difficult to identify sources and the risk of compromising survey objectives exists.

FOCUSING AND SCATTERING EFFECTS

Reflection off a buried surface that contains ridges or troughs can either focus or scatter radar energy, depending on its orientation and the location of the antenna on the ground surface. If a subsurface plane is slanted away from the surface antenna location or is convex upward, most energy will be reflected away from the antenna and no reflection—or a very low amplitude reflection—will be recorded (Fig. 12). This is termed radar scatter. The opposite is true when the buried surface is tipping toward the antenna or is concave upward. In this case, reflected energy will be focused, and a very high amplitude reflection derived from the buried surface would be recorded.

Figure 12 illustrates an archaeological example of the focusing and scattering effects when a narrow buried moat is bounded on one side by a trough and on the other side by a mound. Both convex and concave upward surfaces would be "illuminated" by the radar beam as the antenna is pulled along the ground surface. When the radar antenna is located to the left of the deep moat, some of the reflections are directed back to the surface antenna, but there is still some scattering and a weak reflection will be recorded from the buried surface. When it is located directly over the deep trough, there will be a high degree of scattering and much of the radar energy, especially that which is reflected off the sides of the moat, will be directed away from the surface antenna and lost. This scattering effect will make the narrow moat invisible in GPR surveys. When the antenna is located directly over the wider trough to the right of the moat, there will be some focusing of the radar energy, creating a higher amplitude reflection from this portion of the subsurface interface.

SIGNAL ATTENUATION

Radar signal attenuation with depth is influenced by both the relative dielectric permittivity (RDP), and the electrical conductivity and magnetic permeability of the material through which the radar energy passes (Doolittle and Miller 1991; Duke 1990; Shih and Doolittle 1984). Absorptive attenuation losses of electromagnetic energy increase as the water content of a soil increases and also varies with the amount and types of salts in the medium. High rates of signal attenuation can also be caused by high concentrations of dissolved carbonates within surface soils (Batey 1987). Under the very unfavorable conditions of wet, calcareous, or clay-rich soils, the maximum depth of GPR penetration can be less than a meter, no matter what the frequency of antennas used.

Generally, materials with lower electrical conductivity (high resistivity) allow greater electromagnetic wave propagation and have a low RDP. Materials that have a high electrical conductivity (and therefore a high RDP),

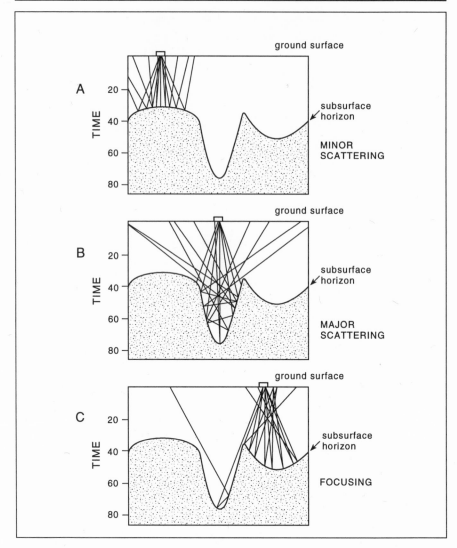

Figure 12. Diagrammatic cross section of a buried surface illustrating the scattering and focusing at different antenna locations. When the antenna is located over a convex upward surface (A), minor scattering occurs. When located over a buried moat-like feature (B), a large degree of scattering occurs because the radar waves are reflected multiple times within the moat and ultimately are scattered away from the surface antenna. When the antenna is located over a concave surface (C), focusing occurs because the radar waves, which are transmitted from the surface in a cone, are reflected back to the surface in a more focused pattern.

such as saturated clays, greatly impede electromagnetic wave propagation; and radar energy is severely attenuated with depth. The attenuation is caused by absorption, due to the conductivity losses in the ground and the spreading out of the energy over a larger surface area with increasing depth (Balanis 1989).

THE NEAR-FIELD EFFECT

Energy radiated from a surface antenna generates an electromagnetic field around the antenna within a radius of about 1.5 wavelengths of the center frequency (Balanis 1989; Engheta et al. 1982; Sheriff 1984). For the 10, 100, and 1000 MHz antennas, this effect is approximately 30 meters, 3 meters, and 30 centimeters, respectively. Within this zone, the radar energy is coupled

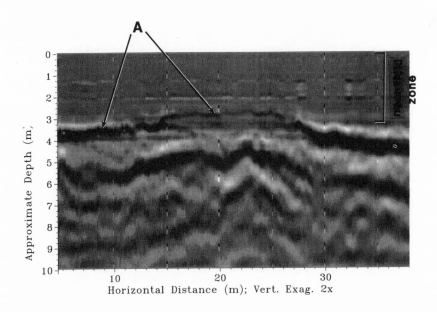

Figure 13. The near-field zone on an 80 MHz profile. The near field zone is about three meters in depth in this profile. The prominent high-amplitude reflection (A) is still visible within this zone, although with a decreased amplitude.

with the ground, generating an advancing wave front in the standard conical transmission pattern *outside* of the radius. It can be said that the ground within about 1.5 wavelengths of a standard dipole antenna is "part of the antenna" in that no radiation is occurring within this zone, and therefore technically no propagation. This near-field zone is usually visible in GPR profiles as an area of few or no reflections (Fig. 13), sometimes called the near-surface zone of interference (Fisher et al. 1992).

If low-frequency antennas are used, the near-field zone where no significant reflections occur can be between 2.5 and 5 meters of the surface. If the features of interest are located within this zone, it is unlikely that they will be visible in GPR profiles and a higher frequency antenna should be used.

There can sometimes, however, be important reflection data recorded within this near-surface zone, even if reflections are not visible on standard two-dimensional profiles. Due to the wide bandwidth of radar transmission, some high-frequency (short-wavelength) energy will still be generated from a lower-frequency antenna. This will couple with the ground at a much shallower depth, and some reflections can still be generated and be visible in profiles. If they are high enough in amplitude, they will appear as weak reflections within the near-field zone (Fig. 13). Others may not be noticeable in standard two-dimensional profiles, but can become visible after the data are computer processed to produce amplitude anomaly slice maps, which are discussed in chapter 8.

Chapter 3

GPR Equipment and Data Gathering

DESCRIPTION OF GPR EQUIPMENT

There are a number of different manufacturers of the GPR units typically used in archaeological surveys. The most commonly used units in North America are manufactured by Geophysical Survey Systems Incorporated (GSSI) in North Salem, New Hampshire, and Sensors and Software of Mississauga, Ontario, Canada. Companies in Europe and Japan also manufacture multiuse GPR systems that have excellent archaeological applications. Most of the systems produced for general purpose GPR surveys employ pulsed radar energy of one center frequency. Other GPR systems employ a stepped GPR technology that focuses a narrow radar beam of varying frequencies into the ground. This technique, which uses a coiled antenna system (Noon et al. 1994), has not had significant archaeological application to date and will not be discussed in this book.

The GSSI and many other manufacturers' antennas are usually pulled or carried across the ground surface, recording data continuously along transects. The Sensors and Software antennas are moved in steps, recording data at specified intervals along lines in grids. Both of these companies are presently marketing GPR systems that can acquire data in either mode because there are advantages to both, depending on the field conditions and types of features to be mapped. There are many similarities and some differences between models of pulse GPR units, and this book will not attempt to go into the details of each system.

Standard GPR systems consist of four main elements—the control unit, the transmitting unit, the receiving unit, and the display unit. The GPR control unit produces a high-voltage electrical pulse, which is sent via cables to the transmitter that amplifies the voltage and shapes the pulse that is then emitted by the antenna. Cables come in varying lengths and are manufactured of either coaxial copper filament or fiber-optic material. Fiber-optic cables, which transmit a digital signal to and from the antennas, are said to reduce greatly some of the equipment-related noise that can affect signal clarity (Davis and Annan 1989). Fiber-optic systems generate the electrical pulse necessary to create radar waves directly at the antenna from self-contained batteries, and only the digitized data from the reflected radar waves are sent through fiber-optic cable back to the control unit.

A standard dipole transmitting antenna consists of a thin copper plate in the shape of a bow tie that radiates the radar pulse into the ground (Kraus 1950). The power (measured in voltage) is applied in pulses to the center of the bow tie. The applied electrical current travels from the center of the antenna to the edges of the copper plate and back again to the middle, creating an electromagnetic field. This radar energy is then radiated from the center of the antenna downward, where it becomes coupled with the ground. The voltage supplied to the antenna can sometimes be adjusted to produce a pulse of greater or lesser power, but in most antenna units the power delivered to the antenna is automatically adjusted for the antenna size and the appropriate frequency by electrical components located within the antenna housing. Too much voltage transmitted to the antenna can overload resistors located at its ends and cause the copper plate to resonate, or "ring," which can severely disrupt the ability of the receiving antenna to record important subsurface reflections due to an abundance of noise in the system. Ringing can also occur if two low-frequency antennas (one transmitter and one receiver) are connected too close together (Sternberg and McGill 1995).

The receiving portion of the antenna records the reflections received on the surface receiving antenna and sends the signal back to the control unit, along a different line located within the cable, to the receiver. The raw reflected signals are then amplified and formatted for display on a video monitor or paper print out. Some systems digitize the signal directly at the receiving antenna before it is transmitted to the control unit along a fiber-optic cable (Annan and Davis 1992). The received signal is then recorded digitally on a computer hard drive or tape. If an older analog GPR system is being used, the received waves, which constitute small voltage changes around a mean, can be recorded on magnetic tape within the control unit for later digitization.

If a single antenna is being used both to send and then receive reflected radar energy, a transducer is used to switch the antenna from a sending mode to a receiving mode and back again. Single antennas are usually necessary only with large, low-frequency antennas—below about 120 MHz—when transporting two antennas to and within the field is a consideration, or when it is important to minimize GPR system ringing.

Paired transmitting and receiving antennas can either be transported along a transect, continuously sending and receiving, or moved in steps. If the two antennas are moved in steps, reflections are recorded at periodic intervals along a transect. With antennas lower than 80 MHz in frequency, data must usually be acquired using the step method because their large size and difficulty of transport make continuous recording inconvenient.

Usually antennas are placed directly on the ground surface, but if they are suspended above the ground it is important that they be located no more than one wavelength off the ground surface in order to increase radar energy penetration (see Table 2). The best antenna position is within a distance of about one quarter of a wavelength of the ground. Inside this distance, the transmitted radar signal will couple with the ground and the best downward transmission will result. If antennas are located too far above the ground, ground coupling may not occur and much of the transmitted radar energy will be reflected back to the receiving antenna from the first interface encountered, which is usually the ground surface. Little radar energy would remain to penetrate into the earth. This same phenomenon can also occur if the antenna is being floated on a freshwater lake, or if data are being acquired through ice and water. In these environments, there will usually be good energy penetration, with moderate signal attenuation, through the water column, but much of the energy may be reflected from the water-sediment interface. Some energy penetration can still occur in the sediment below the water column, but much will have either been attenuated, dispersed in the water, or reflected from the lake bottom–water interface. If significant penetration is required in the lake-bottom sediment, lower frequency antennas may be necessary in order to transmit maximum radar energy below the water column.

Many GPR systems use a hand-held marker device to record the surface position of the antennas along a survey line during data acquisition. When activated, the control unit imposes a noticeable high-amplitude sine wave on analog reflection data covering one or more recorded traces or, on digitally recorded data, identifies bits of data in a trace header. The marker button can either be located on the antenna and operated by the person pulling the antenna sled or attached to the control unit and operated by someone at the command of the antenna operator. It is most common to have the button pushed at standard intervals along the transect, usually every few meters or less, at stations that have been pre-surveyed using a tape measure. When using the step method of data acquisition, the antennas' location on the ground is pre-determined by the step spacing itself, and no marker device is necessary.

RECORDING DATA IN THE FIELD

Reflected GPR data from continuously moving antennas are recorded as a series of discreet waves, called traces, with each trace consisting of the summation of many reflections from ever-increasing depths in the ground. Although the source of the transmitted energy can be thought of as one distinct radar pulse, this is not technically correct. Most GPR systems transmit radar pulses at extremely high rates, ranging from 25,000 to 50,000 pulses per second, and the digitizers in most systems are not fast enough to sample reflected data received from any one distinct pulse. Radar control systems therefore use incremental sampling methods that produce a composite trace by taking the first sample within a trace from a reflection derived from the first transmitted pulse, the second sample from the second pulse and so on until one complete trace is constructed. The digitizer may take 128 or more samples, from the same number of consecutive pulses, to compile the record of one complete trace. Digitizers are now available that are capable of sampling quickly enough to record the reflected data from one distinct pulse, but they are prohibitively expensive and have not yet been incorporated into most commercial GPR units.

In continuous data acquisition, if the antennas are moving across the ground at an average walking speed, incremental sampling will create some averaging of the recorded signal as conditions change in the subsurface. This averaging procedure would affect the recorded data only if the subsurface geology was extremely variable and the antennas were moving at a relatively high rate of speed.

The step method of data acquisition uses the same general procedure, except discrete reflection data are received at every step interval. When the antennas are moved to the next step, reflection data are again acquired. In antennas set up for acquisition in the step method, a beeper sounds or some other form of notice is given after data from one station is acquired to tell the antenna handler to move the antennas to the next station along a pre-surveyed transect. This acquisition method is much slower than the continuous method.

In order to create a cross-sectional display of the subsurface, all recorded traces, no matter what the acquisition method, are displayed in a format in which the two-way travel time of the reflected waves is plotted on the vertical axis and the surface location, or trace number, on the horizontal axis. A number of new techniques for three-dimensional analysis of these same data are discussed later in this book. In standard two-dimensional raw-data profiles produced by moving the antennas continuously across the ground, the horizontal scale is variable because of changes in the speed at which the antennas are moved. The only horizontal scale visible in raw reflection images is that determined by the vertical lines produced at pre-surveyed marker points in analog data (Shih and Doolittle 1984) or by trace headers in digital data. Depending on changes in the speed at which the antennas are pulled along the ground, the number of traces collected between surface marker points would also vary, and the horizontal scale of one continuous profile would be highly variable. Very simple computer processing methods are now available that can correct the horizontal scale, interpolating between the surface marker locations and expanding or contracting the spacing of traces between markers to create a linear horizontal scale. In the step acquisition method, the horizontal scale is set by the distance between acquisition stations and no horizontal adjustment is necessary during post-acquisition processing.

The vertical scale in all GPR profiles is measured in two-way travel time, but it can be converted to approximate depth if the velocity of radar energy in the ground is known. Post-acquisition computer processing typically adjusts both vertical and horizontal scales to create profiles with any desired vertical or horizontal exaggeration. If there is significant topographic variation along a survey line, topographic corrections should also be made that will adjust the subsurface reflections for surface irregularities. This can only be accomplished if detailed surface topographic surveys are made, but it is very important when surface irregularities are prevalent. For instance, if the radar antennas are moved over a low area, subsurface reflections derived from a buried, perfectly flat surface would be traveling for less time through less material and would appear as a bump in radar profiles. The same is true if the antennas were moved over a surface rise. The recorded reflections from the same buried flat surface would then appear as a subsurface trough if not corrected for topography.

ORIENTATION OF SURFACE TRANSECTS

Most archaeological applications necessitate the acquisition of GPR data in a rectangular grid over the area to be surveyed (Doolittle and Miller 1991). If the grid is oriented to the north, then survey lines can be acquired so that there are both north-south and east-west lines. The grid should be situated so that surface obstacles are avoided and located on the most even and horizontal ground possible. If surface obstructions are present, a rectilinear grid pattern with survey lines of different lengths can easily be set up to avoid the obstacles. If buried pipes, tunnels or electrical cables are known to be present, their presence should be identified in advance and the grid should be located so as to avoid them. When this is not possible, the location of these features must be noted so that they can be discriminated from other important reflections when the data are interpreted.

A rectangular or rectilinear grid is preferable to other grid designs for a number of important reasons. Digital reflection data from a rectangular grid can easily be exported to many computer display and imaging programs that are pre-set for this gridding method. With a rectangular grid, the data can be quickly processed and interpreted without time-consuming line surveying and drafting. In addition, each GPR profile in a rectangular grid can be more readily compared to others and reflections can be correlated line by line throughout the grid.

A rectangular or rectilinear grid is not always possible or desirable when conducting a survey. In some cases, surface conditions or time constraints may necessitate a series of separate non-parallel lines that can still yield good subsurface coverage. Care must be taken when setting up a non-rectangular grid so that the location of all lines is accurately surveyed so that the acquired reflection data can be mapped in three dimensions.

All grids must be precisely surveyed using standard surveying techniques either before or after the acquisition of the GPR data. At the very minimum, the corners of each grid must be accurately located, and care taken so that all bounding grid lines are parallel or perpendicular to each other. This part of the GPR acquisition process can sometimes be the most time consuming and tedious part of a survey, but it is extremely important. In the near future it will be possible to survey grids in a cursory fashion and use a global positioning system (GPS) that is either satellite or land-based to record the location of survey lines automatically. With this method, the exact coordinates of each grid line and the elevation of the ground surface could be recorded during acquisition as digital data on a separate channel. This technology, which is becoming more common in geological data acquisition, is just being applied to GPR systems (Czarnowski et al. 1996), so traditional survey methods with a transit and rod, or a laser theodolite, must still be used.

When the ground surface is rough, uneven, or sloping, topographic elevations along each survey line must be obtained so that corrections of subsurface reflections can be made during post-acquisition processing (Sun and Young, 1995). If the ground is evenly sloping it may only be necessary to survey the beginnings, ends, and a few elevations along each line. It may even be possible to obtain only the elevations of each change-of-slope and then interpolate elevations in between to save surveying time. When surface irregularities are numerous, however, elevation surveying must be done at more frequent intervals, usually at all grid-line intersections.

If unshielded antennas that broadcast radar energy in all directions are being used, reflections may be recorded from surface obstacles as well as from the subsurface, so the location of any surface feature that could conceivably reflect radar energy must be mapped. Trees, overhead branches, houses, fences, and overhead power lines must be located and placed accurately on survey maps so that when the reflection data are later processed their reflections can be factored out.

If there is a need to determine the location of subsurface reflection anomalies quickly, with no real interest in a regional map of the site, raw data profiles can be produced and interpreted as the antennas are pulled across the ground randomly. In this case, one could conceivably wander around a site producing reflection records until an anomaly is discovered. Anomaly locations could then be marked on the ground as they are found, and gridding might not be necessary. This is an extremely easy way to conduct a GPR survey, but it is full of pitfalls in most archaeological contexts. For the most part, it is difficult to identify anomalies immediately in raw reflection data. Also, many times the reflections do not appear on the graphic record or computer screen until the antennas are past the anomaly, and then one must estimate its surface location.

In this GPR acquisition method, buried archaeological features can be mapped very quickly, providing the surveyors with instant gratification by being able to visualize the subsurface in a matter of minutes. This approach has its uses, but it should never be used in place of the standard data-acquisition method of conducting surveys in rectangular grids. Also, if the surface location marks are not marked on the records, it can be extremely difficult to analyze or process acquired reflection data later because exact locations of transects are not documented.

ACQUISITION OF SURVEY DATA

Prior to conducting a GPR survey, it is necessary to determine the optimal grid pattern, line spacing, and the antenna frequency to use. Lacking any hard information on the electrical properties of the sediment and soil, estimates must be made as described earlier (Table 2). If excavations, road cuts, or cutbanks are located within the survey or nearby, velocity tests, which will be described in chapter 6, should be performed prior to recording data to make sure that the optimum antenna frequency is being used for the depth and dimension of the target. These types of tests can sometimes be quite time consuming, but they are necessary if accurate maps measured in true depth are to be constructed.

Three people are usually necessary to conduct a GPR survey (Fig. 2), although theoretically one person could do it alone, with lots of running between the control unit, the antennas, and the survey markers. One person is usually assigned to pull the antennas continuously, or move them in steps, along the surveyed grid lines. Another person stays at the control unit to start and stop recording at the beginning and end of each line. If continuous data is being acquired, the person at the control unit can also activate the surface-location marker button by command from the person pulling the sled. In the step method, the control person also tells when it is time to move the antennas. A third person is usually necessary to make sure that any antenna cables do not snag on surface obstacles, help maneuver the antennas when needed, and clear surface debris away from lines.

If the survey grid is laid out in the usual rectangular pattern, then the beginning and end of each survey line must be marked by stakes or push flags. The transect lines can then be marked easily using a tape measure pulled between the stakes or a rope that has been marked with the appropriate grid-spacing interval. Some field workers denote each line and the appropriate survey marks along each line with small pin flags. This method can be quite time-consuming to set up, but is useful if line intersections need to be surveyed later to arrive at topographic correction values.

For continuous data acquisition, once the survey line is conspicuously marked by flags, or with a tape measure or rope, the antennas are pulled along the lines by the antenna puller. This job is the most difficult because the person pulling the antenna sled sometimes must not only walk backward, but also make sure that it is keeping to the designated line. The sled puller must also watch when the middle portion of the sled moves by the designated surface markers and either push the marker button or notify the person holding it when the location is crossed. The sled is usually pulled beside the survey lines so that the flags or markers on the tape measure or rope are visible to the side of the antenna housing. The antennas should always be pulled along the same side of each line for consistency. This will create a minor offset of all lines to one side of the surveyed line locations, but this small error is usually not significant due to the wide conical subsurface transmission pattern.

If antennas are pulled behind a truck or some other vehicle, the same procedure can be used, but with one person walking to the side of the antenna to record the marker locations with the hand button (Fig. 14). Some continuous GPR systems use survey wheels that roll along the ground near the antenna to record surface location data automatically. This method of data acquisition is best used on a very hard, level surface such as a parking lot or paved road. If survey wheels are used on uneven terrain or soft ground, slippage in the wheel can cause sizable errors in the location of marker positions. Some survey wheels pull thread from a spool whose rotation is calibrated to distance. These devices accurately measure distance no matter what ground-surface obstacles may be encountered and can greatly speed up the process.

Figure 14. A single 80 MHz antenna supported behind an oxcart. The person to the side of the cart is operating the surface-location marker button. (Courtesy of Payson Sheets.)

If the antennas are being moved in a step fashion, the same general procedure is used, with the distance between steps constant and the antennas placed directly over each location mark. One or more antenna handlers are still needed to move the antennas from point to point along transects.

Prior to acquiring data along a grid line, the person working the control unit needs to note in a field book the line number in the grid and the corresponding file number to which the reflection data will be saved. This point cannot be understated, because even the most sophisticated computer storage systems can lose data and only a hand-written record will reconstruct the procedures used in the field. If only paper copies of lines are being generated, the control operator should write all the pertinent information about each line on the record so there is no confusion later. The approximate location of obstructions encountered and large trees or other features that could possibly reflect radar energy also need to be documented for each line.

During continuous data acquisition, the cable operator must make sure there is enough cable in reserve to survey each line without stopping. If acquisition is interrupted mid-transect, it is usually best to start over and either record over the old file or record new data in a different file or on a new sheet of paper. If this cannot be done, it is still possible to take line segments from different files and splice them together during post-acquisition processing with data-merging techniques. As the antennas are pulled along the ground, one person must also watch out for any possible snags and tangles. This person can also help move tape measures or survey ropes prior to starting data acquisition on the next line in a grid.

For expediency, during both continuous and step data acquisition, lines are usually surveyed in a sinuous pattern, going forward on one line and then reversing direction on the next. This pattern is continued until all parallel lines in those directions are acquired. Perpendicular lines in the same grid can then be surveyed in the same fashion. If the reflection data are being stored digitally there are simple computer programs that can later reverse all the recorded traces in half of the lines so that all parallel profiles produced within a grid have the same orientation.

EQUIPMENT AND SOFTWARE ADJUSTMENTS

Some manual adjustments are always necessary prior to conducting any GPR survey (Kemerait 1994). These adjustments should usually be made only after estimates or accurate measurements of the velocity of the material through which the radar energy will be passing have been done. The depth to important features that need to be resolved also must be estimated in advance. In newer GPR units, some of the equipment adjustments are controlled automatically by the acquisition software, but these can still be overridden manually. In older models, most of the adjustments must be made manually with switches or knobs located on the control unit.

Header Information

Most digital GPR units have general header information that can be input for each file recorded during a survey. This information typically includes the date of the field work, the antenna frequency, the site name, the grid name or number, and other pertinent information or comments. Most units allow this information to be entered at the start of acquisition, requiring only changes in the line number or other parameters within a grid during the survey. Each line within a grid will usually be saved to the computer hard drive or tape as a separate file. These data must always also be noted in the field book and on paper printouts of the raw reflection data.

Set-up Parameters

Depending on the equipment and software used, there are some basic data that must be entered into the computer in the field before beginning work. Each GPR system has a somewhat different way to do this, but all include the following basic parameters.

Time Window. The time window is defined as the amount of time, measured in nanoseconds, that the receiving antenna will listen to and record the reflected radar-wave energy. The time window is normally opened just before the radar pulse is transmitted and closed after all reflection data of interest have been recorded. If the velocity of the material and the approximate depth of the features to be resolved are known, the amount of time necessary for radar energy to travel down and back can be estimated. The time window can then be adjusted so that it is open for at least this period of time so that all important reflections from all lines in the survey are recorded. The time window should usually be adjusted so that more data, from a greater depth, is being recorded than may seem necessary. Due to unforeseen subsurface velocity changes, reflections from features of interest may be received later than the preliminary calculations estimate; and if the time window is not open long enough, they will not be recorded. It is also possible that reflections of interest may dip to greater depths, or be covered with a greater thickness of overburden along some portions of lines, also necessitating a longer time window.

Samples Per Scan. Once the time window is set on digital units, the number of samples per scan must be set. One sample is a digital value that defines a portion of the reflected wave trace. The more digital samples there are to define a wave, the more accurate the form of the wave becomes. A scan, also called a trace, is a series of reflected waves derived from one transmitted pulse or incrementally sampled from a continuous series of closely spaced pulses. The longer the time window is open, the larger the number of samples necessary to adequately define the trace of the reflected wave. Any number of data samples can be selected per scan on some units, but it is usual to select 128, 256, 512, 1024, or 2048.

This setting can make a large difference in the resolution of important reflections. Usually, the more samples that are recorded, the better the reflected waveform is resolved. If 512 samples are being recorded for each trace and the time window is opened for 512 nanoseconds, then there will be one sample of the reflected waves digitized for every one nanosecond of two-way travel time. This may be more than enough data points to represent a series of reflections. For instance, if the time window is increased to 1024 nanoseconds, then there would be one sample recorded for every two nanoseconds of travel time, decreasing the resolution of the recorded wave. In this case, if the wavelength of an individual reflection of interest is on the order of two nanoseconds, then there would only be one digital sample defining one wave, which is not nearly enough to delineate it. In order to resolve the feature of interest, it would be necessary to increase the samples per scan, or decrease the time window. In most archaeological applications, a time window of 100 to 200 nanoseconds or less is usually sufficient to record reflections within two to three meters of the surface, depending on the material velocity. Therefore 512 samples per scan are usually enough to define the trace adequately.

The maximum resolution that can be obtained is also dependent on the wavelength of the reflected waves that are generated by the antenna, which is a function of the antenna frequency (Table 2). When all these factors are considered, it is easy to see that a number of estimates and assumptions are necessary prior to determining the sampling rate. Some experimentation may be necessary in the field while the antennas are stationary in order to obtain the optimum sampling rate.

It is also important that the time window is not open for too long. The longer the time window is open, more samples per scan will be necessary for good subsurface resolution, and there is a chance too much data will be collected and digitized from depths outside the depth of interest. Storage capacity on a tape or hard drive can become filled quite quickly with useless data, especially if a large survey is being conducted.

Trace Stacking and Recording Rate. The stacking of traces is a method that reduces random and variable portions of the reflected wave. This "noise" is removed by averaging successive traces arithmetically so that one composite trace is recorded (Fisher et al. 1992; Grasmueck 1994; Maijala 1992). This minimizes the variations between reflected traces along a transect that may be caused by interfering FM radio transmissions or by cables, people, or objects that move around near the antenna. Many older GPR units were not capable of stacking during acquisition, and it could only be done during post-acquisition processing, and then only if the data were in digital format. Other older GPR units allow the operator to manually adjust the stack to 2, 4, 8, 16, 32, or more successive traces, and the unit would only record or print out the average trace. Newer models allow any number of traces to be stacked as long as the number is a whole number (Davis and Annan 1989). It is important to recognize that the more traces that are stacked, the slower the antenna must be moved along the ground surface in order to achieve the same horizontal sampling rate and the greatest subsurface coverage.

The stacking process removes anomalous reflections that may be generated from surface irregularities such as small bumps or dips in the ground surface (Fisher et al. 1992; Grasmueck 1994; Maijala 1992). It also filters out the effects of minor velocity changes due to water-saturation changes or small rocks or voids in the subsurface.

Stacking is usually a good idea when the antennas are moving at a fairly slow speed (at an average human pace or less). Stacking eight or more contiguous traces into each one that is recorded can improve the quality of the subsurface reflection data and still give good subsurface coverage. For example, if 80 reflection traces are being measured every second, and each 8 consecutive traces are stacked into one, then 10 traces are recorded each second. If the radar antennas are moving at a rate of 20 centimeters per second along the ground surface, then there is one stacked reflection being recorded for every 2 centimeters of ground covered. If a stack of 16 were applied (each 16 consecutive traces averaged into one) then the subsurface coverage would be one recorded trace every 4 centimeters. Stacking is also used when acquiring data in steps, averaging traces at each data-acquisition location.

Transmission Rate. Radar systems typically transmit at a rate of more than 25,000 pulses per second. With the presently available GPR technology, it is impossible to record each individual reflected trace generated from each transmitted pulse due to the rapidity at which the pulses are being transmitted and then reflected back to the surface. To overcome this problem, radar systems sample incrementally. If the system is set up to stack 16 sequential traces into one recorded trace, then there must be at least 512 pulses times 16 (8192) transmitted for every trace recorded.

An understanding of transmission and recording rate is important for determining the horizontal resolution of the recorded data. Depending on the speed at which the antennas are moving across the ground, adjustments of the recording and stacking rates may be necessary in order to get good subsurface coverage. For instance, if it is necessary to record 4 complete traces per second (after stacking 16 traces into one) then 8192 sequential pulses times 4 (32,768) must be transmitted each second. If the radar device being used is only capable of transmitting at a rate of 25,000 pulses per second, there will not be enough pulses transmitted to allow for 4 recorded traces per second. If this is the case, there are a few minor adjustments that can be made prior to recording data:

- The stacking rate could be lowered.
- The antennas could be moved over the ground slower, recording more total traces per second.
- The time window can be shortened, necessitating fewer samples to define one trace.
- The samples necessary to define one waveform can be decreased.

If the time window is fairly short, the stacking is minimized, and the sampling rate is kept at 512 samples per scan or less, there are usually more than enough pulses being transmitted into the ground in order to allow recording of the necessary traces. The adjustments discussed above are usually required only if the antennas are moving at a high rate of speed (i.e., towed behind a vehicle), there are very high stack rates applied (more than 16 traces stacked into one), or an extremely high waveform resolution (many samples per scan) is necessary in order to define the subsurface reflections. Many recently manufactured GPR units transmit at rates much higher than 25,000 pulses per second, allowing much more latitude in these adjustments.

When collecting data using the step method, there are always more than enough pulses being transmitted to allow for the recording of one trace and the horizontal resolution becomes a function only of the distance between steps.

Signal Position. Prior to acquiring reflection data, it is always important to make sure that the first reflection received from any pulse is located within the time window designated. In digital systems with a computer monitor, the series of pulses and all reflected waves from the subsurface can be viewed on the monitor. Older systems display the trace on an oscilloscope in a similar fashion. The signal position is usually adjusted while the antennas are motionless on the ground surface and radar data are being continuously acquired in one location. The time window can then be manually adjusted so that the first wave (usually the pulse generated from the transmitting antenna) is located at time zero. This first wave is not really time zero because it takes a small amount of time for the pulse to travel from the transmitting antenna to the receiving antenna. If the antennas are separated by more than a few centimeters then a calculation must be done to determine the time it would take for the radar pulse to travel in air between the two. (At the speed of light, which is the velocity of radar in air, the energy travels at .2998 meters per nanosecond.) The signal position can then be adjusted so that the time lag created by the wave moving in air between the two antennas is factored out. If the two antennas are only separated by a few centimeters, the time lag is usually insignificant and the first pulse can assumed to be time zero. Some of the newer GPR units automatically set the signal position during the initial computer setup sequence.

Range Gains. Due to the conical spreading of the transmitted radar waves and the attenuation of the energy as it passes through the ground, later arrivals on a reflection trace will almost always have lower amplitudes than earlier arrivals. To recover this lower amplitude information, gain control is applied during acquisition (Maijala 1992; Shih and Doolittle 1984; Sternberg and McGill 1995) in order to amplify those reflected waves received from deeper in the ground. Gain recovery settings are standard on most GPR equipment; they are automatically adjusted on newer models or can be adjusted manually in others (Fisher et al. 1994; Geophysical Survey Systems, Inc. 1987). There is usually a linear relationship or series of linear relationships between the amount of gain applied and the time at which it is received, with higher gains applied to reflections recorded later. Post-acquisition processing can also be used to boost the signal of some reflection arrivals if the gains applied in the field were not appropriate.

　　When the gain is increased for later reflection arrivals, the amplitude of system noise and other random interference, such as from FM radio transmission, can also be increased. There can also be variations in the amplitudes of reflections along a traverse due to differences in signal attenuation and changes in the antenna coupling with the ground. The largest of these variations can be "smoothed" during data processing by applying automatic horizontal gain recovery to the data (Fisher et al. 1994).

In older GPR systems, gain control knobs are located on the control unit. One knob controls the amplitude of near-time (close to the surface) reflections, and the other far-time waves. Since the reflected waveforms are visible on an oscilloscope, the adjustments can be made to personal preference. Care must then be taken not to touch or otherwise adjust the range-gain knobs during data acquisition so that the same type of data is collected for all the profiles within a grid.

Once gains are set on digital systems, they should remain constant for the whole grid being surveyed with a particular antenna. If they are changed for any reason, the processed reflection data will display differing reflection amplitudes from the same horizons in different parts of the grid, and these may be confused with geologic or archaeological changes of importance. Even if the gain settings are changed and noted during a survey, it may still be possible to normalize the amplitudes of recorded reflections using post-acquisition processing techniques.

Filters Applied During Data Acquisition. One school of thought says that all data should be recorded in the field as raw data and that stacking or other filters should only be applied during post acquisition processing. In this way, all the data, whether good or bad, are acquired. In addition, if the frequency distribution of any one transmitting antenna is not definitively known, important reflections may be generated from frequencies that could unintentionally be filtered out prior to recording. If unfiltered data are acquired, bad data can always be filtered out later during processing.

The other school of thought says that some filtering is necessary during collection so that the data can be more easily interpreted while still in the field. In addition, unfiltered data many times include long-wavelength noise that will affect the gain settings that are applied. Some GPR units can record data on two channels, and therefore both raw and filtered reflection data can be acquired simultaneously on separate channels (Fenner 1992). Both data sets can then be processed back in the office and compared, prior to choosing one or the other for interpretation.

Vertical filters, also called band-pass filters, are used to remove anomalously high- and low-frequency data during recording (Buker et al. 1996;

Fisher et al. 1994). The high-pass filter removes low-frequency data, usually below approximately 10 MHz, that can be generated from "system noise" inherent to each particular radar device. These data can be seen on an oscilloscope or computer display of the recorded traces as long wavelengths superimposed on a standard reflection profile. The amount of low frequency noise will change with the antenna used, the cable length, and the type of control unit and it is usually a function of GPR system design.

Anomalously high-frequency data can be recorded from FM radio transmission or other electrical disturbances nearby. When the antenna is stable and the generated waveform is visible on a computer display or oscilloscope, a short wavelength "flickering" is visible from this high-frequency noise. Low-pass filtering can remove much of this noise, but care must be taken not to remove what may be real reflections. Some newer GPR models will make these adjustments automatically, if desired.

It is important to note that as vertical and horizontal filters are applied in the field, other adjustments such as the time window, sampling rate, transmit rate, and range gains must also be adjusted and possibly reset. These manual adjustments made prior to data acquisition are an iterative process, and a number of experimental profiles should be made prior to gathering data. If good data are being acquired at the necessary depths once the adjustments are set, then the settings should remain the same for all lines acquired within a grid. The computer software included with some digital units allows many of these adjustments to be made automatically, but the reflection data from a few experimental lines should always be viewed and analyzed prior to proceeding with a complete survey.

Chapter 4

Post-Acquisition
Data Processing

The GPR reflection data that is viewed directly after being acquired in the field usually contains what many interpreters call "noise," "reverberations," or "interference"—extraneous reflections that make it difficult to interpret. These "raw" field data are also usually not displayed with corrected depth or horizontal scales. In order to clean up the noise and correct the horizontal and vertical scales of the raw data, they must be processed prior to interpretation.

There are a large number of commercial and proprietary GPR processing programs available for GPR data. Many of the data processing techniques in them have been borrowed and modified from the petroleum industry, which processes seismic reflection data, and from remote sensing applications, which deal with complex image processing (Malagodi et al. 1996; Milligan and Atkin 1993). To do most post-acquisition processing, data must either have been recorded digitally or be analog data that has been digitized.

The simplest processing procedure takes the individual reflection traces and prints them in sequential order so that they may be viewed as a two-dimensional vertical profile through the ground. These can be printed in "wiggle-trace" format, which shows the individual traces and their amplitudes, or in a grey scale, where amplitudes of individual reflections vary in shades of grey. Profiles can be exaggerated in the vertical or horizontal direction to emphasize certain aspects of the stratigraphy or buried archaeological features.

Standard wiggle-trace or grey-scale profiles can also be modified so that the relative amplitudes of reflections are assigned colors. In this way, significant reflections that may represent important interfaces in the ground are readily visible. Care must be taken in choosing a color palette, however, because sometimes many-colored sections can be "busy" and difficult to interpret (Milligan and Atkin 1993).

More complicated processing techniques that remove portions of the reflection record in order to enhance other aspects can be applied to standard two-dimensional profiles. Some sophisticated techniques attempt to "collapse" portions of some reflections in order to transform into two-dimensions what is really a complex three-dimensional reflection pattern caused by the wide cone of radar transmission. As with all computer modification techniques, alterations of the data should only be applied for specific reasons and may not be warranted for all data sets. One should never attempt to use processing programs "off the shelf" without understanding the implications of each data manipulation technique.

BACKGROUND-REMOVING FILTERS

The most common type of filtering that can be applied to any digital data set, either saved to disk or on tape, is the removal of the horizontal banding that appears in many GPR records. Due to the "ringing" of some antennas, horizontal bands (Fig. 15) are recorded on most profiles (Shih and Doolittle 1984; Sternberg and McGill 1995). These bands can obscure reflection data that would otherwise be visible on some profiles. Horizontal bands can also represent reflections from objects, such as the person pulling the antenna sled, that were a constant distance away from the antenna during acquisition. Most processing programs have the ability to remove these bands in a simple arithmetic process that sums all the amplitudes of reflections that were recorded at the same time along a profile and divides by the number of traces summed. The resulting composite digital wave, which is an average of all background noise, is then subtracted from the data set. The post-filtering profiles then will display only non-horizontal reflections, or those horizontal reflections that are short in length (Fig. 15).

Care must be taken when applying a background-removing filter to areas where subsurface stratigraphy is horizontal, or nearly so. If this processing

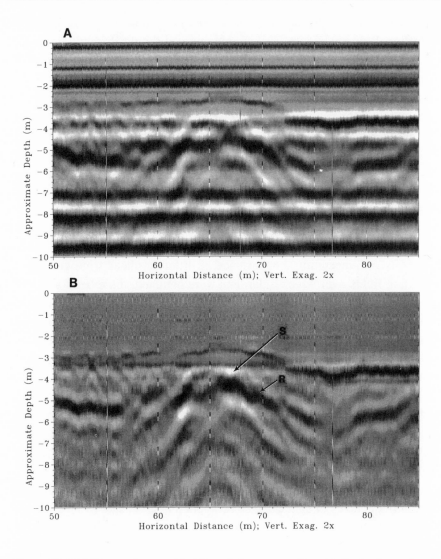

Figure 15. 80 MHz raw data profile (A), and the same profile with the background noise removed (B). The raw data in (A) exhibits horizontal banding due to system noise and the recording of reflections from objects a constant distance away from the surface antenna. The lower image shows the same section after the horizontal bands have been removed (B) with a background removal program. (S) is the location of a buried structure at the Ceren Site in El Salvador. (R) is a point-source reflection generated from the buried clay floor of the structure.

step is used on GPR profiles from such areas, most, if not all, of the important reflection data may be lost. Background removal should also only be applied on a digital reflection data set with a sufficient number of traces. The background removal filtering process removes all the reflection events that occur at the same time, leaving only those that are more random. Those random reflections usually represent significant archaeological or geological reflections in the subsurface, but if too few reflection traces have been acquired over too small a distance, the subsurface geology and archaeology may not have changed enough to have created non-horizontal reflections. The averaging involved in removing the background would then remove most of the important data as well as the noise, leaving little reflection data of any sort.

Background-removal filters are also very useful in survey situations when the ground surface over which a GPR grid is acquired consists of variable soils and vegetation. For instance, if different plants were growing in adjoining fields over which radar data were obtained, the reflection signals obtained from the two fields could possibly vary. Profiles obtained in recently plowed land could also vary considerably with respect to their background signal strength from those obtained over ground that is planted or lies fallow. Similarly, differing background strengths can occur between profiles that are acquired parallel to row crops as opposed to perpendicular. When this happens, certain portions of the reflected data may show differences in reflection strengths due to differences in ground-coupling caused by variations within surface soils. By applying background filters to all lines collected in a grid, the relative strengths of reflections can sometimes be normalized between different areas of a site.

Other more sophisticated background-removing filters that remove only certain frequencies of data can also be applied to GPR data (Malagodi et al. 1996). This type of processing can remove extraneous reflections that are often referred to as clutter or noise.

F-K FILTERS

There are many experimental processing techniques, originally developed for seismic data processing by the petroleum exploration industry, that have been applied to GPR reflection data (Lehmann et al. 1996; Maijala 1992;

Milligan and Atkin 1993; Yu et al. 1996). Care must be taken when applying these techniques, because there are some important differences between radar and seismic reflection data.

F-k filtering is a technique where reflections recorded in time are transformed into frequency data using complex arithmetic transform programs (Maijala 1992). The outcome of this processing technique is that high-angle reflections (possibly also point-source reflection hyperbolas) that may be obscuring important horizontal data are removed. This seismic data-processing technique has been misapplied to some GPR data and should only be used with caution.

DECONVOLUTION

Deconvolution is another seismic-reflection processing technique that has been applied to GPR data (Fisher et al. 1994; LaFleche et al. 1991; Maijala 1992; Malagodi et al. 1996; Neves et al. 1996; Rees and Glover 1992; Todoeschuck et al. 1992). It is based on the theory that, as a radar pulse is transmitted into the ground, portions of the electromagnetic wave will change form, or convolve. The purpose of this filter is to remove the portion of the recorded waves that have convolved during transmission. Deconvolution processing restores the reflected waves in a profile to their original pattern and presents the data with a different look. In this fashion, the deconvolution technique can be used to identify and resolve some hard to identify reflections.

One of the problems with deconvolution processing is that restoring reflected waves to their original forms is mostly a function of educated guesswork. It is difficult to determine what the original waves really looked like, and any deconvolution process may be modifying the data in unreal ways. If deconvolution processing of radar data can be improved, it may yield important clues to understanding how the earth modifies transmitted electromagnetic waves and help in GPR interpretation. Most attempted applications of this processing step have so far proved to be unsatisfactory (Maijala 1992; Rees and Glover 1992).

MIGRATION

Standard GPR systems portray a distorted image of subsurface stratigraphy, and features caused by both the wide beam of radar propagation and changes in velocity with depth. Migration is a two-dimensional imaging process that has been used with limited success to eliminate some of the distortions caused in data collection procedures (Fisher et al. 1992, 1994; Grasmueck 1994; Malagodi et al. 1996; Milligan and Atkin 1993; Young and Jingsheng 1994). Distortions that are caused by the wide beam of radar antennas generate reflections from point sources that appear as hyperbolas. Migration processing will focus and "collapse" the reflection hyperbolas back to the point at which they were reflected in the subsurface. Although this processing technique has so far been used with limited success in GPR (Maijala 1992), its applications could be expanded in the future as computer processing of data becomes more routine.

Migration is not a standard technique in most GPR processing programs and can be time-consuming and expensive. At present, the effects of the wide cone of transmission on the migration of hyperbolas has not yet been addressed in the literature. There are some seismic processing programs available that have been customized to accommodate GPR data, but difficulties still remain when interpreting GPR data as if they were seismic data (Milligan and Atkin 1993). In addition, most migrations of reflections in seismic data are in two-dimensions, and radar data are always acquired in three-dimensions. Three-dimensional seismic migration programs exist, but their application can be time-consuming and their applicability to GPR data remains unknown.

In many cases, it may be better to view radar profiles in an unmigrated format because point-source hyperbolas are many times capable of identifying subsurface anomalies that represent archaeological features of interest. If all the hyperbolas are collapsed back to the point of origin during processing, the features that generated them may be more difficult to recognize, causing important anomalies to go unidentified. The opposite could also be true if important reflections were generated within the area where point-source hyperbolas are recorded. In this case, hyperbolas that are migrated back to their point of origin would allow these important reflections to become visible.

Chapter 5

Synthetic GPR Modeling

One of the most confusing aspects of GPR for archaeologists is that the two-dimensional profiles produced by moving antennas along transects do not always "look like" the stratigraphy or archaeological features they are used to seeing in trench profiles or outcrops. For instance, a small buried house floor in a GPR profile might appear as a hyperbola, or a steep-sided trench as an X shape. To overcome this problem, archaeologists have recently used computer-generated synthetic radargrams as a way to model buried objects, stratigraphy, and important reflection surfaces in two dimensions.

Such modeling provides an idea of what real-world GPR reflection data "should look like" and can allow a more accurate interpretation of GPR profiles once they are processed (Annan and Chua 1992; Cai and McMechan 1994; Goodman 1994; Goodman and Nishimura 1993). It also allows the construction a model of the known stratigraphy and archaeological features to determine, prior to going to the field, if a GPR survey will be capable of delineating the features of interest. Once models are constructed, they can be modified quickly for different frequency antennas to determine the optimum equipment to take to the field. After GPR data have been acquired in the field and processed, models can be readjusted to represent known field conditions more accurately. When used in this way, they are a great benefit in interpretation, especially when features whose origin is not immediately known are visible in GPR profiles.

Computer-simulated radargrams are generated by tracing the theoretical paths of radar waves during transmission and reflection through various media with specific relative dielectric permittivities (RDPs) and electrical conductivities. The two-dimensional geometry of the subsurface stratigraphy and archaeological features are programmed into the model to generate as close to a real-life case as possible.

It is often the case that GPR processed reflection profiles printed in two dimensions look significantly different from the buried structures or other archaeological features being searched for. They are not at all like images such as those from X rays in medical technology that most of us are used to seeing. One of the reasons for this is that GPR antennas transmit energy into the ground in a wide beam; the antenna is therefore not only looking straight down, but also in front, in back, and to the sides. For example, when the antenna is in front of a buried object, the travel time for a wave to leave the antenna, reflect off the object, and return directly to the antenna is longer than when the antenna is directly over the object (Fig. 3). The net effect creates a hyperbolic reflection pattern over the object as the antenna moves over it (Fig. 16). Large buried rocks, burials, void spaces, or other archaeological features that are buried in the ground can act as similar "point sources" for reflection and will often generate a hyperbolic-type reflection pattern. Depending on the electrical properties of layers within the ground, the frequency of the radar energy being used, and the amount of velocity change at the interface with the point source, reflection hyperbolas can vary dramatically in their geometry. Point source reflections can many times be confused with an upward bowing of material or some other stratigraphic change in the overburden material (e.g., Loker 1983). Using synthetic radargram modeling, known archaeological features and surrounding stratigraphy can be imaged in advance, and their origin can easily be identified, simplifying interpretation.

Another reason that GPR reflection profiles can look very different from the actual features that need to be resolved is the phenomenon of multiple reflections. Multiple reflections can often occur within the ground as radar waves are reflected more than once off subsurface discontinuities before being recorded at the surface. Usually most of the radar energy that is reflected at a subsurface interface between two materials is transmitted directly back to the surface and recorded at the receiving antenna. Some of that reflected energy, however, can be re-reflected back into the subsurface at the ground surface–air interface, and then re-reflected back again to the surface from the same buried

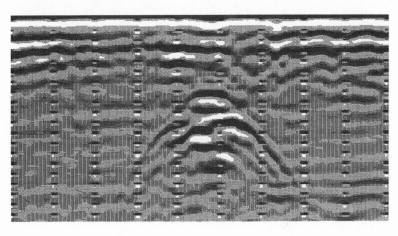

Figure 16. A buried tunnel (upper photo) that was crossed by a radar profile (lower figure). The tunnel is represented by a hyperbolic reflection pattern, with the apex of the hyperbola denoting the top of the tunnel (from Goodman and Nishimura, 1993). The tunnel was discovered at the Kofun 111 burial mound in Saitobaru Park, Japan.

interface prior to recording. This multiple reflection effect creates a double "echo" of the subsurface reflector, with a recorded time about twice that measured for a wave reflected only once. The multiple reflection, which appears about twice as deep in the ground in two-dimensional profiles, is usually much lower in amplitude due to geometric spreading during its travel, energy attenuation, and additional reflection from numerous other interfaces along its path.

The synthetic radargram technique can create a model of important parts of an archaeological site only if some prior information about the site is available. The stratigraphic and electrical characteristics of sediment and soil conditions also need to be inferred, as do the geometry of overburden units and the archaeological features of interest. These data are then put into the computer to create a two-dimensional model that is a simplification of a slice through the earth. The computer can use this information to predict the reflectivity coefficients encountered at various interfaces, the signal attenuation with depth, the velocity of radar energy in different units, and the amplitude of received reflections (Goodman 1994). After the model is run, resulting reflections are plotted in two dimensions in the same fashion as standard GPR profiles. When displayed in this fashion, the relative amplitudes of reflections can be highlighted by using different grey scales or color palettes.

During interpretation, profiles from real field data can be directly compared to the profiles generated by the computer model. Input parameters can then be changed, and the simulation can be re-run until a reasonable match between the real and synthetic profiles is obtained. This iterative process of parameter input and comparison to real-world data is often referred to in geophysical prospecting as "forward modeling" (Powers and Olhoeft 1995) and is a powerful interpretation tool.

CREATING A SYNTHETIC RADARGRAM

In order to generate a synthetic radar profile, large numbers of potential radar wave paths through the two-dimensional model that approximate the paths rays would take in the field are calculated on a computer. This type of two-dimensional modeling is well known in seismic processing, and it is often referred to as "ray tracing" in geophysical terminology. In GPR modeling, unlike seismic modeling, the conical-shaped transmission patterns of the surface transmitting antennas must be accounted for (Goodman 1994). To accomplish this, ray-path models that account for conical spreading and the complex conductive and dissipative nature of electromagnetic wave propagation in the ground must be calculated on the computer. This process can require many thousands of calculations per model, depending on the number of geological layers and archaeological features to be accounted for. The reflected amplitude responses of the radar waves are then predicted, and a two-dimensional GPR simulation is created.

Possible wave paths that must be considered when creating a synthetic radargram will be examined for the model shown in figure 17. There are four different media here in which radar waves can travel from the surface antenna: air, Soil Unit 1, Soil Unit 2, and bedrock. A synthetic radargram computed for this simple four-layer stratigraphic model creates computer-synthesized radar waves at a surface antenna, sends them into the modeled ground, and shows them partially reflected off some interfaces and partially transmitted across others. Partially transmitted rays are also modeled that are reflected off the deeper interfaces. The computer predicts the time at which the modeled reflected waves will return to the receiving antenna and predicts their resulting amplitude. The amplitude and direction of each starting wave is set by the program, depending on the antenna frequency and the RDP given for the units.

In this simple model, some of the simulated radar waves will reflect off the interface between Soil Units 1 and 2 and will be recorded back at the surface antenna. Other waves will be transmitted through this interface and be reflected off the interface between Soil Unit 2 and the bedrock. In the real world, only a portion of the energy from each radar pulse would be reflected and refracted at each boundary, but for simplicity the model is set so that each radar ray has one distinct path, some examples of which follow.

- One wave will be reflected off the interface between Soil Units 1 and 2, traveling directly back to the surface where it is recorded at the receiving antenna (Fig. 17a).
- Another wave travels the same path, but instead of being recorded at the receiving antenna, it is re-reflected back into the ground from the air–Soil Unit 1 interface. The wave paths of these multiple reflec-

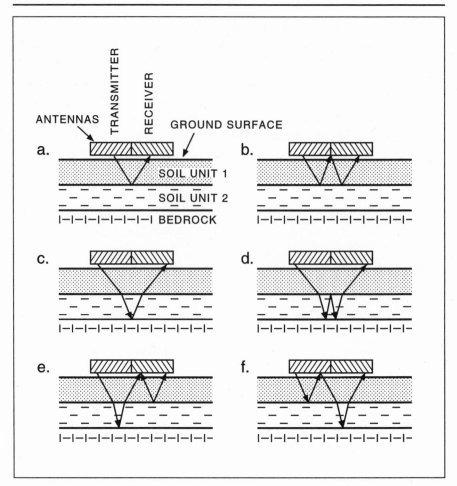

Figure 17. Models of radar transmission and reflection in a three-layer medium. In (a) a single reflection occurs between Soil Units 1 and 2. A multiple reflection occurs at the same buried interface and at the ground surface in (b). In (c) there is refraction at the interface between Soil Units 1 and 2, and reflection occurs at the interface of Soil Unit 2 and the underlying bedrock. In (d), (e), and (f), there are multiple refractions and reflections off of subsurface interfaces prior to the wave's being recorded at the surface antenna.

tions will then travel back into the ground to be re-reflected back to the surface at the same interface (Fig. 17b).

- A similar ray path with multiple reflections is also simulated at the interface between Soil Unit 2 and the bedrock (Fig. 17c).
- There will also be ray paths simulated that take what are called "dog-leg" paths, traveling in complicated multiple reflection paths off of all the modeled interfaces prior to finally arriving back at the surface (Figs. 17d, 17e, 17f).

These, and other paths within the cone of illumination of the transmitting antenna, are simulated on the computer many thousands of times to arrive at a composite of all the reflected waves that arrive back at the surface antenna. The computer then moves the antenna along the modeled ground surface and the process is repeated for many surface antenna locations along the transect.

To compute the synthetic profile, all the modeled reflected energy that is returned to the receiving antenna from each of the many hundreds of paths is totaled at each antenna location over the profile. The time at which each reflection is received and the resulting amplitudes are also recorded, and the simulated reflection traces are plotted in a standard two-dimensional profile. The number of reflections or transmissions that occurred at each interface can be adjusted by the modeler. Using the RDP and electrical conductivity data that are input for each simulated stratigraphic unit, the computer calculates the reflection coefficients at each interface and the amount of radar energy that is reflected or transmitted at each. Attenuation along the path of each modeled wave is also computed from the relative dielectric permittivity and electrical conductivity estimates that are programmed into the computer.

In the real world, there are an infinite number of possible wave travel paths and resulting amplitudes, but if the model is constructed without too much complexity, large numbers of possible ray paths will generate a coherent and usable synthetic model. The following is a sequence of processes involved in the construction of a simple two-dimensional synthetic radargram calculation:

1. Send a computer-generated synthetic radar ray into the model from the ground surface that has a specific traveling itinerary of reflec-

tion, transmission, or both. These must be inferred by the person constructing the model based on a knowledge of the subsurface stratigraphy and archaeological features.

2. As the synthetic ray travels in the different media, the computer will integrate the distance traveled and continually compute the attenuation and resulting amplitude of the wave. The distance and velocity changes along its path are also continually computed in order to measure travel time.

3. When the radar wave strikes a subsurface interface the computer computes the change in amplitude due to either reflection or transmission. It will also estimate the angle of refraction at the interface due to changes in velocity along the interfaces.

4. If the ray returns to the surface antenna, its travel time and remaining amplitude are recorded. Rays that are scattered away from the surface antenna are not recorded.

5. Steps 1 to 4 are then computed for every angle or subangle radar rays might take within the surface antenna's cone of illumination, depending on their travel paths. These must be computed because the strength of a synthetic wave is stronger directly beneath the antenna than to the front and back, creating reflection amplitude variations.

6. Steps 1 to 5 are then computed for all the important combinations of various ray paths within the cone of illumination for all locations along the two-dimensional transect being modeled.

All of these steps are calculated in the computer program only within a two-dimensional plane. The net result of the six steps above is a two-dimensional synthetic radargram. Details about the electromagnetic theory involved and the mathematical description of all steps in the model is given in Goodman (1994).

In real field conditions, there are numerous out-of-plane reflections that also must be taken into account, but to date only two-dimensional models have been constructed for GPR. Three-dimensional modeling techniques have recently been developed for seismic reflection data used in petroleum exploration, but these have not yet been adapted for GPR.

SYNTHETIC RADARGRAM APPLICATIONS IN ARCHAEOLOGY

In areas where buried features have considerable geometric variation within a short distance, GPR reflection profiles and the synthetic radargrams that model these features can be very complex. An example of a synthetic radargram for a buried V-shaped trench with steep sides and two different subsurface layers is shown in plate 1a. In this model, which simulates a constructed moat around a small ceremonial or burial site, a material with a higher RDP overlies one with a lower RDP. For simplicity, the electrical conductivity of both layers was assumed to be the same. The model predicts that some radar waves will have a single reflection off the subsurface interface, while others are reflected twice or three times within the trench before returning to the surface. The direct reflected waves (those with only one reflection off the interface) show the outline of the trench, with less amplitude on the edges of the trench due to ray scattering. Because the antenna will "see" the far wall of the trench before encountering it, due to the conical transmission pattern from the surface antenna, reflections from the trench side will be recorded (in measured time) at a depth greater than the actual trench bottom. The same is true as the antenna moves away from the trench, when reflections are received from its opposite side. The net affect of these reflections, as the antennas are moved along the ground surface, causes what is referred to as a "bow-tie" effect. Multiple reflections within the trench also create a hyperbola directly under the bow-tie feature. Other multiple-reflection features are also created when radar waves are reflected three times, but these are recorded as very low amplitude waves due to progressive energy attenuation during transmission. In real reflection profiles, triple multiples of this sort, which are predicted by the computer to be recorded near the ground surface and below the bow tie, would probably not be visible due to their very low amplitude.

In the V-trench case shown in plate 1a, if the synthetic model was not constructed and analyzed, the rounded feature beneath the trench might be mistaken for a point-source hyperbola derived from something buried below the trench. The model, however, demonstrates that these reflections on the profile are the result of multiple reflections from inside the trench and do not represent a real feature.

A similar situation is shown in plate 1b, where a more rounded trench contains a void space (modeled with a RDP of 1) within it. In this model, the void space represents a burial within the trench. The same physical characteristics of the overburden material and the reflection wave types were used as in the V-shaped trench example (Pl. 1a). The resulting synthetic model shows a high amplitude point-source hyperbolic reflection, with long axes projecting down to the base of the synthetic radargram, that is generated from the velocity discontinuity at the void space. The upper edge of the trench is shown as a strong reflection, with its relatively flat bottom distorted due to a bow-tie effect similar to that visible in the V-trench example.

This more complicated model in plate 1b illustrates how the void space would be discovered easily in a real-world situation by its visibility as a point-source hyperbolic reflection. The upper edges of the trench would also be visible by direct reflections, but its base would be obscured or distorted due to the bow-tie effect and the interference of the point-source reflection hyperbola that was generated from the void. Reflections recorded on the profile that were generated from multiple subsurface reflections are barely visible due to their low amplitude, and only tend to obscure and complicate other more important reflections in the synthetic radargram.

A model that illustrates how layered materials of varying thickness, with differing RDP and electrical conductivity, produce reflections is shown in plate 1c. In this model, the flat interface between Units 2 and 3 represents an ancient living surface that is the archaeological interface of interest. The complexity in this model arises from the differences in RDP and varying thickness of the overlying two units. The modeled changes in RDP directly affect the velocity of radar energy that passes through the two materials that overlie the flat ancient living surface.

The thick section of material with a high RDP will slow the radar energy as it travels vertically from the ground surface to the interface of interest and back to the ground surface. The thinner section of this high-RDP material will allow the radar energy to pass from the ground surface to the interface of interest and back to the ground surface in a shorter amount of time because it is traveling a shorter distance in the lower velocity material. The resulting reflections generated from the buried interface of interest, when plotted in two-way travel time, will be distorted due to these velocity and thickness differences. Under the area where a thinner section of low-velocity (higher RDP) material is found, the surface will appear to bow upward. The opposite is true under the area where the low-velocity material is thicker. This upward and downward bowing of the interface of interest, caused only by differences in the velocity and thickness of the overlying material, creates the illusion of an undulating surface between Units 2 and 3, referred to as a velocity "pull-up" or "pull-down."

A velocity pull-up similar to the model in plate 1c could also be found below a large buried void space. The increase in radar wave velocity within the void would create an artificial upward bowing of those reflections produced below the void. A pull-down could also occur where localized conditions create a decrease in radar wave velocity, possibly due to abrupt stratigraphic or archaeological changes, or to changes in surface soil conditions. A fluctuating water table, or changes in the water saturation of buried units located above the water table, can also slow radar waves and distort underlying reflections due to localized conditions.

In the model in plate 1c, the undulating contact between Units 1 and 2 located above the interface of interest, is accurately portrayed by the upper reflection in the synthetic radargram; only the interface of interest below these units is distorted. In a real-world situation, if the RDPs of the two upper layers were known, the velocity changes could be accounted for and the distortion of underlying reflections could be adjusted during data processing. This model demonstrates one of many pitfalls that interpreters of GPR data can encounter when there are large changes in velocity within the overburden material. In addition to these problems, the radar waves, because of complex refraction and transmission within subsurface layers, may not be transmitted through and therefore not reflected from some buried features. These "un-illuminated" regions, also referred to as "shadow zones" (Goodman, 1994), are areas where no reflections occur, even though features that could reflect energy are present.

SYNTHETIC RADARGRAMS COMPARED TO GPR PROFILES

Synthetic radargrams were created at the Ceren Site in El Salvador in order to model an area around and over a buried structure. The mid–Classic-age Ceren site is a Mayan village that was almost instantaneously buried by volcanic ash about A.D. 590 (Sheets 1992). Numerous structures and the buried living surface were almost perfectly preserved under more than five meters of volcanic ash and other ejecta. A structure consisting of a raised house platform built on the buried ground surface was computer modeled (Pl. 2a). In this synthetic model, the relative dielectric permittivities of the volcanic overburden, the ancient ground surface, the construction material used in the structures, and a clay layer under the ancient living surface were obtained both from field tests and from laboratory measurements of samples collected in the field (Conyers 1995a). The model was constructed to simulate a 300 MHz center frequency antenna.

A RDP of 3.2 was used for overlying volcanic units from the surface to a depth of two meters. At a depth of two meters, the RDP was abruptly increased from 3.2 to 5, although, in the real world, the RDP likely increases gradually with depth due to a gradual water saturation increase. For simplicity in the model, this variation was not taken into account in order to make the synthetic reflections as uncomplicated as possible.

The synthetic radargram that was created assumed a large change in RDP from 5 to 12 at the ancient ground surface. This assumption was based on field observations that documented an increase in water saturation and clay content at the interface and also from direct velocity measurements that are described in chapter 6. A structure with two standing columns or walls was computer rendered on a topographic rise in the middle of the model, based on excavations of structures that were carried out nearby (Kievit 1994). The structure was "covered" in the model by about five meters of volcanic ash. The predicted character of the reflections in the resulting model could be compared directly to real GPR profiles that were obtained over known structures nearby. This was possible because GPR profiles had been acquired over the structures prior to their excavation.

(continued on page 105)

Color Plates

Plate 1a. Synthetic two-dimensional computer model of a buried V-shaped trench. A change in relative dielectric permittivity is modeled between the two materials, which generate single and multiple reflections at the interface. The outline of the trench can be seen in the simulation with a "bow tie" effect created directly beneath the trench. The bow tie effect is created by the wide field of view inherent in radar antennas.

Plate 1b. Synthetic two-dimensional computer model of a rounded trench with a buried void. This model simulates a burial chamber or cache within a trench. Relative dielectric permittivities are the same as in plate 1A. A reflection hyperbola created from the void space (with a RDP of 1) is created. The upper edges of the trench are visible in the reflection model, but its base is obscured due to the bow tie effect, similar to that created in the V-trench model in plate 1A.

Plate 1c. Synthetic two-dimensional computer model of a three-layer system with undulating layers. The relative thickness differences in the layers with RDPs of 7 and 15 create velocity variations within the two-dimensional model. The changing velocities of radar transmission make the lower horizon appear to bow upward in the reflection simulation, where the overlying lower-velocity material is thin. The opposite is true when the overlying lower-velocity material is thicker.

Plate 2a. Synthetic two-dimensional computer simulation of a buried house platform at the Ceren Site, El Salvador. The relative dielectric permittivities of units were derived from laboratory and field tests. The computer-derived reflection patterns show reflection hyperbolas generated from structure walls and the floor of the platform. A high-amplitude reflection is also received from the interface of the buried living surface and the overlying material.

Plate 2b. Three-dimensional color representation of the buried topography at the Ceren Site as mapped by GPR. The view is from the southeast at 20 degrees above the horizontal. The buried topography is vertically exaggerated two times.

Plate 3a. Three-dimensional view of figure 30 after color rendering. Buildings, trees, the central plaza, and other anthropogenic features are rendered on the computer using the digital information displayed in figure 30. (Image courtesy of Fenton-Kerr Engineering.)

Plate 3b. Amplitude anomaly slice map constructed from 60 to 80 nanoseconds (3.9 to 5.2 meters depth) at the Ceren Site, El Salvador. The brightly colored red and orange anomalies represent the location of buried structures that were constructed on the buried living surface. These anomalies represent the location of high-amplitude reflections generated at the interfaces of clay walls and floors with the matrix material, which is volcanic ash. This map was derived from the reflection data in Grid 4, shown on figure 39.

Plate 4. A series of horizontal amplitude slice maps of the Nyutabaru Site, Japan. The buried circular moat and central chamber are most clearly visible in the slice from 40 to 48 nanoseconds. In the lowest slice, a possible fence and corral are visible as linear anomalies to the west of the circular moat.

Plate 5a. Time-slice maps of the Kofun 111 burial mound, Saitobaru Park, Japan. The surface moats and rings are clearly visible in slices from the surface to 42 nanoseconds. In the slices from 56 to 98 nanoseconds, interior features, which may represent burial chambers and a tunnel leading to them, are visible. Additional interior features, deeper in the mound, are visible in the slices from 112 to 168 nanoseconds.

Plate 5b. Time-slice maps of the Yamashiro Futagozuka Mound GPR grid, Japan. The burial chamber is visible as a brightly colored anomaly in the eastern portion of the grid within the slices from 40 to 70 nanoseconds.

Plate 6a. Three-dimensional cutaway of the GPR anomaly visible in the slices shown in plate 5B that was created by the burial chamber, Yamashiro Futagozuka Mound, Japan.

Plate 6b. Horizontal amplitude anomaly maps constructed from the topographically corrected, two-dimensional profiles, Spiro Mound 6, Oklahoma. The slice from 35 to 42 nanoseconds images a rectangular feature that may represent a wooden or earthen structure within the mound. (After Goodman et al. 1995.)

Plate 7a. Amplitude time-slice maps produced at the Matsuzaki Site, Japan. Amplitude anomalies in the slices from 21 through 42 nanoseconds represent the location of archaeological features. A modern garbage pit is visible in all slices while the older archaeological features are visible only in the deeper slices. (After Goodman et al. 1995.)

Plate 7b. Amplitude slice maps, Shawnee Creek Site, Missouri. A number of high-amplitude anomalies are visible in many of the deeper slices that represent buried dwellings and other unexcavated archaeological features.

Plate 8. Amplitude time-slices, uncorrected for topography, over the Kofun Mounds 102 and 103, Japan. The slices from 0 to16 nanoseconds show surface features and the location of the mounds. In the slice from 32 to 40 nanoseconds, vertical tunnels are visible as high amplitude red and yellow anomalies. Possible burial chambers in Mound 102 are visible in the deepest slices from 40 to 64 nanoseconds.

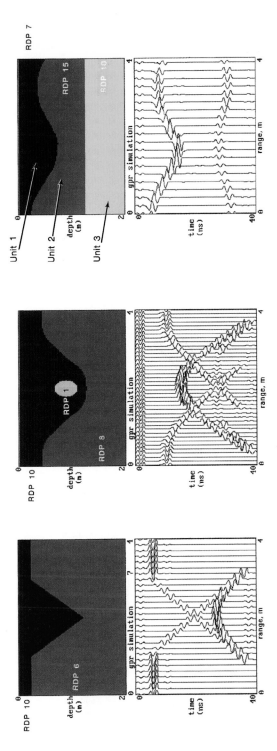

Plate 1. Synthetic two-dimensional computer models of:

Plate 1a. Buried V-shaped trench.

Plate 1b. Rounded trench with a buried void.

Plate 1c. Three-layer system with undulating layers.

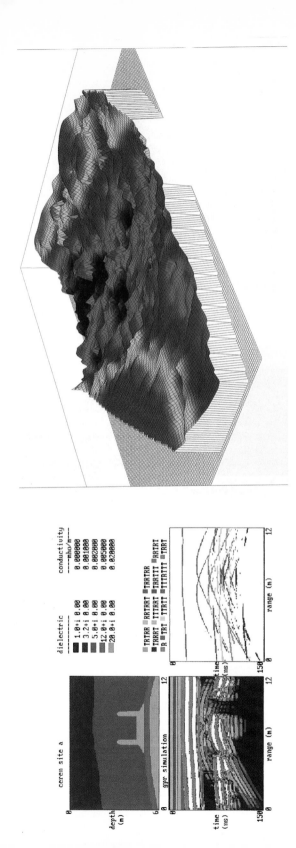

Plate 2a. Synthetic two-dimensional computer simulation of a buried house platform, Ceren site.

Plate 2b. Three-dimensional color representation of buried topography, Ceren site.

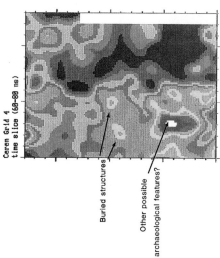

Ceren Grid 4
time slice (60-80 ns)

Buried structures

Other possible
archaeological features?

N ←

Plate 3a. Three-dimensional view of figure 30 after color rendering.

Plate 3b. Amplitude anomaly slice map constructed from 60 to 80 nanoseconds, Ceren Site.

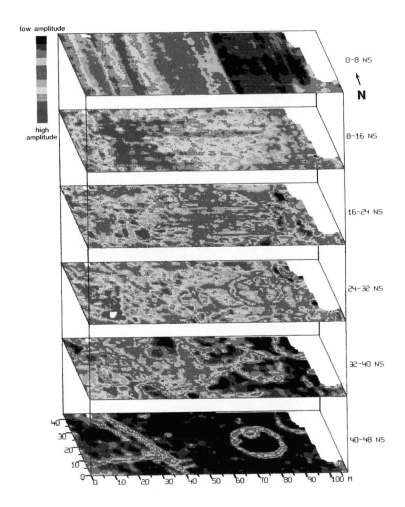

Plate 4. A series of horizontal amplitude slice maps, Nyutabaru Site, Japan.

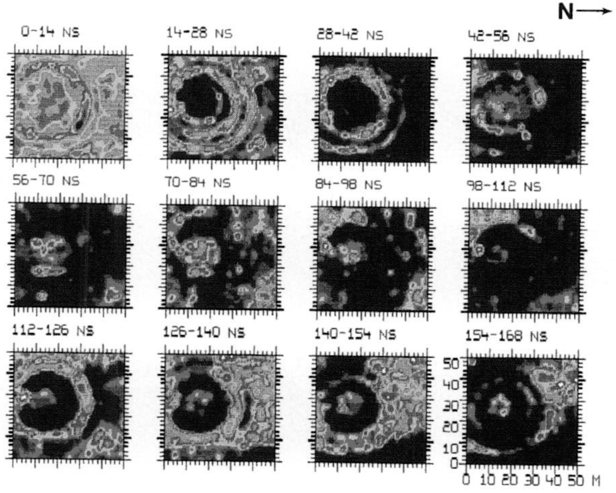

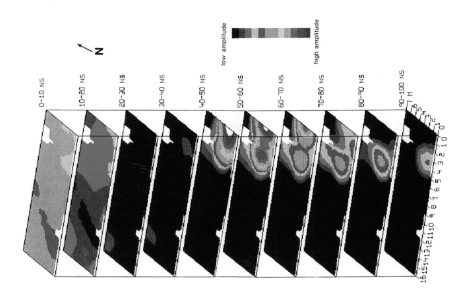

Plate 5a. Time-slice maps of the Kofun 111 burial mound, Saitobaru Park, Japan.

Plate 5b. Time-slice maps of the Yamashiro Futagozuka Mound, Japan.

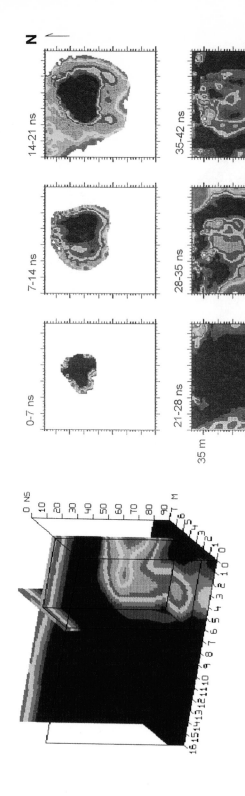

Plate 6a. Three-dimensional cutaway of the GPR anomaly visible in the slices shown in plate 5B.

Plate 6b. Horizontal amplitude anomaly maps constructed from topographically corrected, two-dimensional profiles, Spiro Mound 6, Oklahoma.

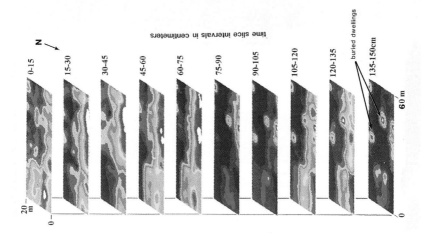

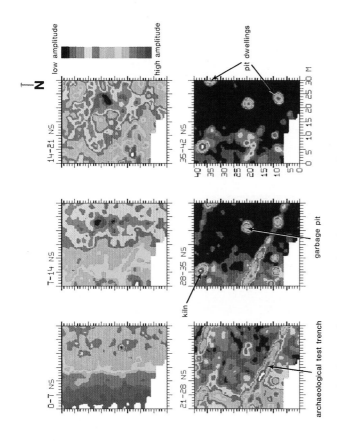

Plate 7a. Amplitude time-slice maps produced at the Matsuzaki site, Japan.

Plate 7b. Amplitude slice maps, Shawnee Creek site, Missouri.

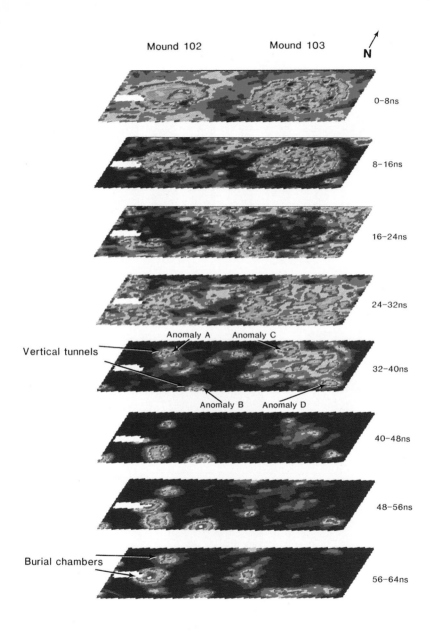

Mound 102   Mound 103   N

0–8ns

8–16ns

16–24ns

24–32ns

Anomaly A   Anomaly C

Vertical tunnels

32–40ns

Anomaly B   Anomaly D

40–48ns

48–56ns

Burial chambers

56–64ns

Plate 8. Amplitude time-slices, uncorrected for topography, over Kofun Mounds 102 and 103, Japan.

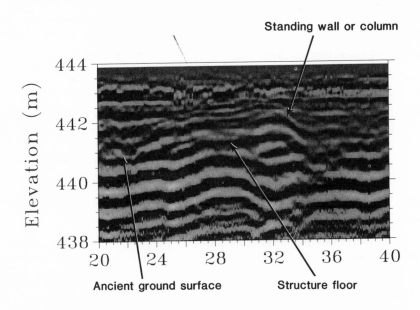

Figure 18. A 300 MHz GPR profile across a buried platform structure with standing walls or columns, similar to that modeled in plate 2a. Point-source reflections occur from the tops of the walls or columns, as predicted. The structure's floor and the ancient ground surface are also visible as high amplitude reflections, as predicted in the synthetic model.

In the synthetic model shown in plate 2a, point-source reflections were generated from wall or column tops. The apex of these hyperbolas is plotted within the volcanic overburden. This same phenomena is documented in a number of real 300 MHz profiles where the transects cross a buried structure with standing walls or columns (Fig. 18). In this case, reflections were generated from the top of the columns or walls and from the floor. The synthetic radargram also predicted a high-amplitude reflection would be generated at the contact between the ancient living surface and the overlying volcanic ash. This predicted response replicates what is seen in many GPR sections at sites where the ancient living surface has been identified and the structures are known to exist.

The computer-generated models of the buried structures and living surface were used during interpretation as a basis for identifying both the ancient living surface and the structures built upon it (Conyers 1995b). Other interpretation techniques that partially employed data derived from the synthetic models are discussed in chapter 7.

The construction of synthetic two-dimensional radargrams adds an important new dimension to GPR data interpretation. In real-world conditions, stratigraphic changes and the dimensions of archaeological features can be highly variable and may be difficult to interpret in GPR profiles. Reflections can also be altered due unforeseen subsurface water saturation changes and soil conditions. These are only some of the variables that must be accounted for in real-world conditions. Without the use of modeling programs that can simulate these types of variations, many important buried features may go unrecognized or be misinterpreted.

The use of synthetic radargram modeling prior to going to the field can also be of benefit in selecting the equipment to be used. If field conditions are known, or can be estimated, the correct antenna frequency can be chosen and optimum line spacing within a grid determined. The ability to construct a number of synthetic models quickly allows many possible variables in physical properties and archaeological feature geometry to be accounted for.

Chapter 6

Time-Depth Analyses

One of the primary purposes of modern GPR surveys is to map stratigraphy and buried archaeological features accurately. In the past, many GPR studies, especially those conducted in archaeological investigations, had the limited objective of finding buried anomalies that might represent archaeological features that could be excavated later. The actual depth and orientation of the features discovered and the nature of surrounding stratigraphy that may have been related to those features was usually of secondary interest. In contrast to most of the early "anomaly hunting" GPR studies, many recent surveys have been conducted for the purpose of non-invasively mapping buried features in detail, sometimes without ever having to excavate. Many times, when excavations are planned as a follow-up to geophysical mapping, GPR maps can very accurately delineate specific areas to concentrate on without the need for extensive exploratory digging. To accomplish these goals, precise subsurface mapping in true depth is necessary. Archaeologists with a geological background have also learned that GPR data yield excellent stratigraphic information about the sediments and soils that surround excavations that are rarely analyzed using standard archaeological techniques. Stratigraphic information, unobtainable in another fashion, can be of great value when reconstructing buried topography, studying anthropogenic disturbance, and analyzing post-depositional processes that may have affected an archaeological site.

This changing focus of GPR exploration in archaeology has necessitated accurate subsurface mapping in real depth. Specific reflections visible in GPR data, which are always measured in two-way travel time, must be tied to known stratigraphy or archaeological features at measurable depths. This conversion of two-way travel time to depth must be done before any realistic data interpretation can begin. Time-depth conversions can only be established if the velocity of the material through which the radar energy is traveling can be calculated. This chapter describes a number of field and laboratory tests that can be performed to arrive at these velocity measurements.

Radar-wave travel time is the only direct measurement that can be obtained using GPR equipment in the field. Depth (or distance) can be directly measured using a measuring tape. If both time and distance are known, velocity can be calculated. There are two general field techniques for determining velocity: the reflected wave method and the direct wave method. Reflected-wave methods require that radar waves be reflected from objects or stratigraphic interfaces at depths that can be directly measured (Sternberg and McGill 1995). Direct-wave methods transmit the radar waves directly through the ground, from one antenna to another along a measured distance. Both these methods determine velocity by measuring the time it takes radar waves to travel known distances.

Multiple velocity tests should be conducted in a study area because it is common for the velocity of overburden material to change laterally as well as with depth. Spatial velocity variations are most commonly caused by changes in water saturation and lithology. Water content is usually the single most significant variable that affects radar wave velocity. Dry quartz sand has a RDP of about 4, which calculates to a radar wave velocity of 14.99 cm/ns. In contrast the RDP of pure water is about 80, which yields a radar velocity of 3.35 cm/ns. If only a small amount of water is contained in the pore spaces of dry sand, the velocity of radar energy traveling in it will decrease significantly. In most settings, the water content of soil and sediment will naturally increase with depth, and the average radar wave velocity of the material will correspondingly decrease. The degree of residual water content in sediment and soils located above the water table, as well as the depth to the water table, can fluctuate dramatically across an area due to changes in surface topography, stratigraphy, and the location of drainage features. In archaeological contexts, buried anthropogenic features can also create layers that affect water saturation and therefore create dramatic velocity differences across a site. Velocity is not only influenced by water saturation differences, but also by changes in lithology of sediment and soils. Many times it is difficult to determine the causes of velocity differences across an area because they can be related to both water saturation changes and lithologic differences, or both.

It is important to recognize that velocity measurements at a site are valid only for GPR data that are collected within a few days of when the tests are performed. Changes in velocity can vary dramatically with the seasons as sediment and soil moisture fluctuates; velocity can sometimes change rapidly, even during the time a survey is being carried out, due to torrential rainfall, snowmelt, or flooding. For example, velocity tests performed at the Ceren site in El Salvador during the rainy season yielded a RDP of 12 (average velocity of 8.7 cm/ns)(Doolittle and Miller 1992), while tests performed in the same area at the end of a six-month dry season measured a RDP of about 5, or an average velocity of 13.4 cm/ns (Conyers and Lucius 1996). In this case, if the velocity tests performed during one season were used to process and interpret GPR data acquired just a few months later, the calculated depths of significant radar reflections would be extremely inaccurate.

REFLECTED-WAVE METHODS

The most accurate and straightforward method to measure velocity is to identify reflections in GPR profiles that are caused by objects, artifacts, or zones of interest that occur at known depths. These methods allow for a direct determination of the *average* velocity of radar waves from the surface antenna to a measured depth. In the past, these types of velocity tests have been conducted at archaeological sites on objects as diverse as buried whale bones (Vaughan 1986), copper wire (Kenyon 1977), and empty paint cans (Doolittle and Miller 1992).

The reflected wave tests that were performed at the Ceren site are used here as an example of this method (Conyers and Lucius 1996). They are referred to as the "bar test," the "wall test," and the "stratigraphic correlation test." The bar test involved pounding an iron concrete-reinforcing bar into the side of an excavation and identifying it on a GPR profile. Because metal objects are near-perfect reflectors, the reflections generated from them are easily identifiable on most GPR profiles. The wall test was analogous to the bar test, except the top of a buried adobe wall, an end of which had been exposed in an excavation, was identified on GPR profiles. The stratigraphic correlation test incorporated the velocity measurements derived from the bar and wall tests to help identify reflections generated from important stratigraphic horizons exposed in test pits.

The Bar Test

In this test, an iron bar was pounded into the side of an excavation exactly 1.1 meters below the ground surface. Dual 500 MHz antenna were then placed at the ground surface and the antennas were slowly pulled over the bar while subsurface reflections were recorded. In order to obtain the maximum amount of reflection from a thin metal, bar the long axis of the surface antennas must be oriented parallel to the length of the bar. This antenna orientation will create an electric field that is oriented parallel to the bar, also producing the maximum amount of reflection. If the antenna axes were oriented perpendicular to the bar, only a small portion of the transmitted radar energy would be reflected from the bar, and it would probably not be visible in profiles. Care must also be taken to use an antenna frequency suitable for the depth and object size that needs to be illuminated. If a low frequency antenna is used, a large object is necessary in order for it to be visible on standard two-dimensional profiles. For instance, the 2 1/2-inch diameter iron bar used at the Ceren site was not visible at a depth of 1.1 meters using 300 MHz antennas but was with the higher resolution 500 MHz antennas.

During the field velocity tests, it was disconcerting that the iron bar was not visible in the raw reflection displays. Only later after the data were processed using a background removal program, discussed in chapter 4, was the bar visible (Fig. 19) as a reflection hyperbola, with its apex denoting the top of the bar.

In this test, the two-way travel time from the surface to the bar and back to the surface was measured at 13 nanoseconds, and the measured depth to the bar was 1.1 meters. Equation 1 can then be used to solve for the relative dielectric permittivity (K) of the material between the surface and a depth of

Apex of reflection hyperbola

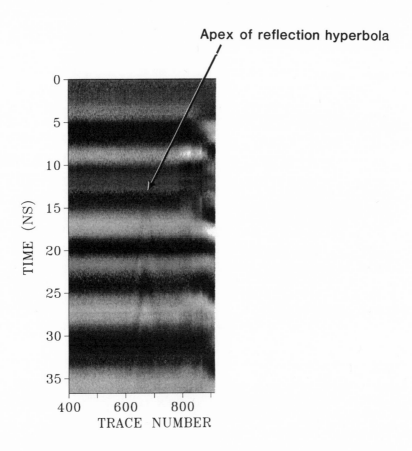

Figure 19. Reflection hyperbola generated by a metal bar. The bar was buried 110 cm below the ground surface. The reflection hyperbola apex is visible at 13 nanoseconds (6.5 ns one-way time).

1.1 meters. It is important to remember that the measured two-way travel time must be divided by two in order to arrive at the one-way travel time, prior to establishing the velocity (time divided by distance) of the radar waves as they travel from the surface to the buried feature.

$$\sqrt{K} = \frac{.2998 \times 13/2}{1.1}$$

K = 3.14

The bar-test calculation derived a relative dielectric permittivity from the ground surface through the upper portions of the overburden material of 3.14. It should be noted that these surface layers are composed of volcanic ash that was almost devoid of clay and were extremely dry during this test, which was performed at the end of the dry season. As noted above, a similar test performed in about the same area during the rainy season yielded a RDP of 12, illustrating the large effect of water on radar-wave velocity. Using these same time and depth measurements, the average velocity of radar waves from the surface to 1.1 meters can be calculated as 16.92 centimeters per nanosecond (110 cm/6.5 ns), which is a very high velocity for GPR waves.

Sometimes it is possible to conduct a variation of the bar test if a point source hyperbola of unknown origin is visible on a profile while field acquisition is still being carried out. If the anomaly's surface location is noted, it can be uncovered, its origin determined, and its depth measured. The same velocity calculations performed for the bar test can then be made and used to correct time to depth during data interpretation.

The Wall Test

Another time-depth test was performed near the bar test, where the corner of a buried structure with standing walls had been partially excavated. A two-meter-long portion of the structure's adobe wall could be projected into the overburden where the test was performed. The top of the structure's wall was measured at 2.51 meters below the ground surface. A number of GPR profiles were then acquired perpendicular and parallel to the wall using the 300 MHz antennas. The wall was not visible in those transects that were oriented parallel to the wall because the long axes of the antennas were perpendicular to the orientation of the wall and little energy was reflected. The wall was visible only on those profiles oriented perpendicular to the buried wall. Other test lines were located nearby, but not over the structure, so that a representative profile that did not contain reflections from the buried wall could be obtained for comparison. The two most diagnostic GPR profiles are shown in figure 20.

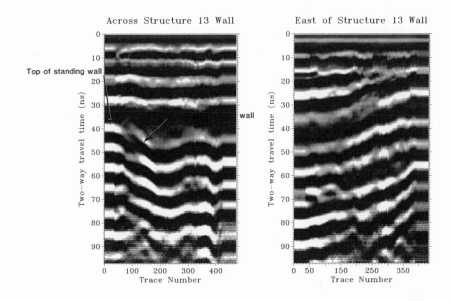

Figure 20. GPR profiles of the wall test. The left profile was acquired perpendicular to a buried wall, the top of which was measured 251 cm below the ground surface. The top of the wall is visible in profiles at 38 nanoseconds, with an ash dune along its north edge. The right profile was acquired two meters to the east of the buried building, where only a regional dip of the volcanic stratigraphy is visible in the reflections.

In the profile that crossed the wall the top of a volcanic ash unit can be seen creating a dune to the north of the standing wall. This ash dune was also visible in exposed faces along the sides of the excavation. The top of the wall is visible in the GPR section that crossed the wall at 38 nanoseconds. The radar profile acquired to the east of the structure demonstrates only a regional dip of the volcanic beds where no dunes had formed. Using Equation 1 in the same way as the bar test, where the known depth to the wall is 2.51 meters and two-way travel time is 38 nanoseconds, calculates a RDP from the ground surface to the wall of 5.15. The time and depth obtained from this test yield an average velocity of 13.21 centimeters per nanosecond (251 cm/19 ns) from the ground surface to a depth of 2.51 meters, a slower average velocity than that calculated in the bar test due probably to an increased water saturation with depth.

The results of these two time-depth calculations are shown in table 3.

Table 3
Velocity Analysis Results from Bar and Wall Tests

Depth (meters)	Total Distance (cm.)	Unit No.	Two-way Time (ns.)	RDP	Velocity (cm./ns.)
0–1.1	220	13–8	13	3.14	16.92
0–2.51	502	13–3	38	5.15	13.21

The most obvious conclusion that can be drawn from these two tests is that the average velocity near the ground surface is greater than at depths greater than about a meter. When using these data to convert radar travel time to depth during post-acquisition processing, care must be taken to use the correct RDP value. Reflections created from zones of interest located only in the upper meter or so should be converted to depth using a RDP of about 3. However, if the reflections to be mapped were located below about 1.5 meters, a RDP of about 5 should be used.

If a RDP of 5 were used to process all the GPR profiles in a nearby grid, then only those reflections below about 1.5 meters would be accurately placed in depth. If this were done, then those reflections from buried interfaces higher in the stratigraphic section would have been converted to depth using a RDP that is too high (and a corresponding velocity that is too low), and they would appear in profiles as too deep. The alternate would be true for the deeper reflections if a RDP of 3 were used in data processing.

This is one of the pitfalls encountered using an average RDP to correct time to depth during processing. If different RDP values were known for specific depths, it would be possible to convert profiles measured in time to depth profiles using varying RDPs for different depths. This approach to time-depth conversion would still create some distortion in the resulting profiles at the boundaries between RDP values because abrupt velocity changes would be inferred that may not be real. A more accurate approach would be to obtain RDP values at varying depths in the section and compile a RDP variation curve for the stratigraphic section as a whole. A mathematical equation that defines the varying RDP with depth could then be applied to reflection data measured in time during processing in order to create more accurate depth profiles. This more complicated approach has not yet been commonly applied to GPR time-depth conversions in archaeological studies.

Stratigraphic Correlations

Once accurate estimates of velocity at differing depths are obtained from direct measurements, a comparison of reflections visible in profiles to known stratigraphy can be made. At the Ceren site, a stratigraphic correlation test was performed on east-west GPR profiles bounded on either end by recently excavated test pits (Conyers and Lucius 1996). A complete section of the important stratigraphy at the site was exposed in both pits that could then be correlated to GPR reflections in the profiles.

The ancient living surface at Ceren, which was buried by the volcanic eruption, is called the *tierra blanca joven* (young white earth), or *TBJ* for short. The uppermost few centimeters of the TBJ formed the surface on which people walked, structures were built, and crops were grown. Because of rapid burial by the volcanic eruption, this surface was almost perfectly preserved under the overlying volcanic ash and coarser ejecta (Sheets 1992). Fifteen different volcanic units were deposited above the TBJ surface, reaching a total thickness of more than five meters in some locations (Miller 1989). The volcanic units overlying the TBJ consist of alternating fine- and coarse-grained pyroclastic materials that have a very similar chemistry but differ in density, porosity, and residual water saturation. All fifteen volcanic units and the TBJ were exposed in the test pits on either end of the GPR profiles. For this reason, two GPR transects were located so as to intersect the excavations on either side; this would allow direct correlations of GPR reflections to stratigraphy on either end of the profiles.

The 500 MHz antennas were used to collect reflection data along one of the stratigraphic correlation lines to image the shallow stratigraphy from the ground surface to a depth of about 2.5 meters, which is approximately 30 nanoseconds in two-way travel time (Fig. 21). Three hundred MHz antennas were used along the same traverse to resolve stratigraphy to about 5 meters, or approximately 80 nanoseconds in two-way radar travel time. GPR data were acquired up to the edges of each excavation on both traverses so that the resulting reflections could be directly tied to the known stratigraphy in pits at either end.

The most-continuous and highest-amplitude reflection on the test line was measured at 62 nanoseconds (two-way time), on the western edge of the line (Fig. 21). This extremely strong and laterally continuous reflector was probably caused by a large velocity contrast between units of different lithology with a related change in their water content. Calculating the approximate depth of the reflection on the west end of the test line, using an average velocity of 13.2 cm/ns that was obtained from the wall test, arrived at an approximate depth of 409 centimeters below the surface.

In the western test pit, the first dramatic lithologic discontinuity that could conceivably have caused this strong reflection is the contact between the top of the TBJ living surface and the overlying volcanic material, which was measured at a depth of 4.2 meters. Recalculating the velocity, assuming that the high amplitude reflection at 62 nanoseconds represents the top of the TBJ, derived an average velocity through the volcanic overburden to the top of the TBJ of 13.55 centimeters per nanosecond, only a little slower than that calculated for the wall test (Table 3). Subsequent interpretation of more than 7600 meters of reflection data within five nearby GPR grids confirmed that this reflection was in fact generated at the top of the TBJ. The buried TBJ topographic surface, as defined by GPR profiles that were converted to depth, almost perfectly corresponds to the orientation of buried topography exposed in four large excavations and fourteen nearby test pits (Conyers 1995b). The computer-generated synthetic models discussed in chapter 5 also predicted a strong reflector at the top of the TBJ, further confirming this correlation.

It is also possible to verify the velocity calculations from the bar and wall tests by correlating other stratigraphic layers to reflections visible above the TBJ in the stratigraphic test profiles (Fig. 21). The 500 MHz reflection profile was capable of resolving specific units within the volcanic overburden to a depth of about 2.5 meters. In order to confirm the velocity calculations from the shallower bar test, the strong reflector, visible at 13 nanoseconds on the eastern

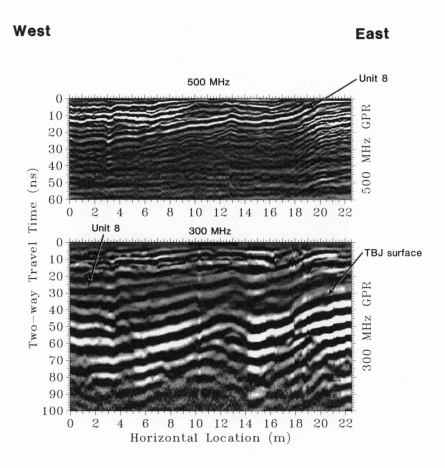

Figure 21. 500 and 300 MHz profiles, both acquired along the same surface transect. Reflections were correlated to stratigraphic layers at measured depths on either end of the profiles. The high-amplitude TBJ reflection is correlative along the profile in the 300 MHz data. Shallower volcanic units such as Unit 8 are also correlative using both the 300 and 500 MHz data.

edge of the line, was assumed to represent the top of a hard volcanic unit visible in the test pit. This unit, informally termed the *capa dura,* or "hard layer" at the site, is numbered Unit 8 (Miller 1989). It is the densest volcanic unit and therefore holds the most interstitial water of any of the uppermost volcanic units. Observations in the test excavations on either end of the correlation line indicated that the abrupt change in water saturation and induration between Unit 8 and the overlying less-saturated volcanic units would probably create a strong reflection in the GPR profile.

To test this assumption, an average velocity of 16.92 cm/ns, which was calculated from the bar test (Table 3), was used to calculate the depth of this strong reflection. Using this velocity, the change in lithology that produced this reflection would occur at 109.98 centimeters deep in the pit on the east end of the line (16.92 x 13/2). In this pit, Unit 8 was measured at 136 centimeters below the surface, close to what was calculated, lending some corroboration to the shallow velocity measurements derived from the bar test. To further test the accuracy of the Unit 8 correlation, the reflection was correlated to the west along the test profiles. Unfortunately, it was not possible to correlate it using solely the 500 MHz reflection data because the unit dips from east to west and the reflection descends below the depth of resolution for 500 MHz data. Fortunately the reflection is also visible, although with less precision, in the 300 MHz reflection profile in figure 21. On the west end of the line, the hypothesized Unit 8 reflection occurs at 35 nanoseconds. Unit 8 was measured at exactly 231 centimeters below the surface in this western pit, which calculates an average velocity of 13.2 centimeters per nanosecond (231 cm x 35 ns/2). This is exactly the velocity calculated from the wall test (Table 3), corroborating the Unit 8 correlation and also the shallower velocity calculation from the bar test.

These series of stratigraphic tests along the test lines yield a high degree of confidence, at least in the area of the site where the tests were performed, that two way travel time of radar waves can be accurately used to calculate the true vertical depth of reflections. Using these velocity measurements, an average RDP for the overburden material was then calculated with equation 1 and used to convert time to depth in all grid lines.

It is quite possible that the imposition of only one RDP, derived from tests that were performed in only one area of the site, on all of the GPR data acquired in the surrounding grids could yield spurious depth calculations if subsurface conditions changed in other areas. For instance, greater depths of the TBJ in some areas would theoretically allow for a greater amount of water-saturated volcanic overburden, lowering the average velocity of waves traveling to and from the surface. The TBJ would therefore appear in GPR profiles to be too shallow because a higher average velocity (and lower RDP) was assumed. Conversely areas where the TBJ is located in a higher-than-average paleotopographic position would have less volcanic overburden and a correspondingly lower overall water saturation. A more geographically dispersed velocity calibration data set would allow more accurate time-depth conversions throughout the area and would overcome these problems. Without such calibrations, these potential velocity problems must be accepted as part of the inherent imprecision in the GPR method.

DIRECT-WAVE METHODS

Although not as accurate as reflected-wave methods, direct-wave techniques provide an additional way to determine radar-wave velocity in the field. In these types of tests, the two antennas are separated, with the material to be tested between the two. One antenna then transmits to the other, and the one-way transmit time between the two can be measured. If the distance between the two antennas is known, velocity can be calculated.

The transillumination method projects radar energy between two antennas that are separated by the material to be measured. This can usually be done only if the material is preserved between two nearby excavations or outcrops. To perform these tests, the transmitting antenna is placed along the exposed face in one excavation, while the receiving antenna points toward it along a parallel face in the nearby excavation. In this way, radar energy is directed in a straight line between the two, and the distance between the antennas can be directly measured. This method has not commonly been used in archaeological velocity tests, but has been used with success in testing the integrity of concrete or stone pillars (Bernabini et al. 1994).

Other similar types of velocity tests are commonly called common midpoint (CMP) (Fisher et al. 1994; Malagodi et al. 1994; Tillard and Dubois 1995), radar surface arrival detection (Hanninen et al. 1992) or wide-angle reflection and refraction (Imai et al. 1987; Milligan and Atkin 1993). These types of tests place two antennas on the ground surface; as radar waves are transmitted between the two, they are either separated from a common point or begin apart and are moved together. Radar energy that is sent from one will pass through the air and near-surface layers and be received at the other. If the distance of separation is known and the radar-wave travel paths can be deduced, the arriving waves can be measured in time and a series of velocity measurements of different layers can be calculated.

In both the transillumination and common midpoint-type tests, one-way radar travel time is being measured. The two antennas that are necessary to conduct these types of tests need not be of the same center frequency because the wide band-width of commonly used GPR dipole antennas will still allow them to "communicate." For instance, a 500 MHz center frequency antenna, which transmits radar energy in frequencies ranging from approximately 250 to 1000 MHz, can be used in conjunction with 300 MHz antenna, which ranges in frequency from 150 to 600 MHz, because their band-widths overlap over a fairly large range. It is best, however, to use antennas that are as close as possible in center frequency. An 80 MHz antenna could theoretically transmit to a 900 MHz antenna, but much less of the transmitted energy would be detected than if the two were closer in center frequency. Dual antennas that are already mounted together for regular GPR acquisition can be used for these tests without physically separating the two. When this is done, only one antenna in each unit is activated, one to send and the other to receive. In order to perform these tests, two cables—one to transmit and one to send—and a GPR system with two channels can be used. The same type of configuration can also be accomplished on a single channel unit if a cable splitter is used that divides the transmission and reception lines into two separate cables.

Transillumination Method

The transillumination method is widely applicable to archaeological settings because there are commonly two nearby excavations where the section of material to be tested is exposed. The faces of the excavations or outcrops should be as close to parallel as possible. It is best if tests are performed soon after the material is exposed in the excavations so that any artificial evaporation or seepage of water along the faces does not significantly change the water saturation characteristics of the material.

Two antennas, one to send and the other to receive, are then placed on the walls of the two excavations, pointing toward one another (Fig. 22). It is important that the two excavations be separated by at least one-and-a-half wavelengths of the center frequency of the antenna being used to transmit. If they are closer than this, the receiving antenna may be within the near-field zone of the transmitting antenna and the first received signal in the resulting profiles may be difficult to identify.

A series of transillumination tests can be made starting at the base of the excavations and moving upward. The two antennas can be moved upward either in steps or continuously as radar energy is transmitted between the two. If they are moved continuously, care must be taken to keep the antennas a known distance apart and a known height from the base of the excavation. If the antennas are moved in steps, it is important that each antenna be moved the same distance from the top or bottom of the exposed faces so that the distance is known between the two. If the walls of the excavation are sloping, then a series of distance measurements must be made in order to arrive at the antenna separations.

When the material to be tested is highly layered, it is important that the electric field generated by the dipole antenna be oriented parallel to the bedding. To do this, the long axes of antennas must be placed perpendicular to the bedding planes. In this way the electrical portion of the electromagnetic field will vibrate parallel to the layers, and there will be maximum isolation

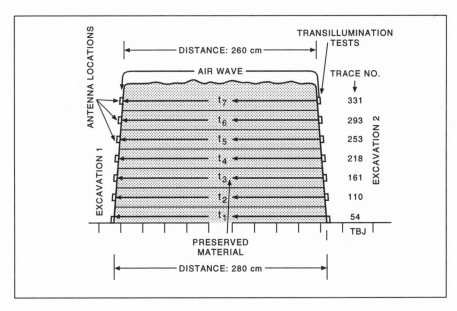

Figure 22. Transillumination test set up. One-way travel times (t_1–t_7) were measured in seven steps performed along two facing excavations, with the preserved material being tested between. The arrival times of air waves that may have traveled around the top of the excavation were also calculated.

of the radar waves within each layer. The elongated cone of illumination of the radar antenna may, however, transmit radar energy into adjacent layers regardless of the orientation of the antennas. If the material to be tested is fairly homogeneous and unlayered, the orientation of the antennas is not as important as long as the transmitting and receiving antennas are oriented in the same direction so that there is maximum communication between them.

The transillumination method was employed at the Ceren site in an excavation where an exposed section of volcanic ash was preserved between two excavations (Fig. 22). The eight different ash units and the buried TBJ living surface were exposed along the sides of the excavations and were transilluminated in seven steps, from the base upward. The 500 MHz antenna was put on one side of the exposure and the 300 MHz antenna on the other, with the 500 MHz antenna used to transmit and the 300 MHz to receive. The horizontal distance between the two antennas at the top of the preserved section of volcanic material was measured at 2.6 meters; due to the sloping walls of the excavations, it was approximately 2.8 meters at the base.

Seven transillumination measurements were made in steps from the base upward. The first step in the test transmitted radar waves through the lowest part of the exposed ash and the TBJ, which appeared to be more moist than the overlying units. Six additional tests were performed on specific ash units above the lower test, moving the antennas in steps to within about 30 centimeters of the surface.

The visual display of the received waves and representative traces at each step in the transillumination test is shown in figure 23. In the grey scale reflection profile, it is difficult to measure the time of arrival of the first reflection precisely, although the arrivals are roughly visible as the first change in the grey scale color. To measure the first arrivals more precisely, representative scans from each step were overlaid on the profile, and the first wave arrival was measured exactly using a trace display program (Powers and Olhoeft 1994). The first arrival in these scans is usually identifiable as the first significant change in amplitude. The period of little or no response in each trace is the time that elapsed between when the pulse was sent by the transmitting antenna and when it was received at the antenna on the other side of the exposed material.

In the upper five steps, very high amplitude reflections were received after the first arrival that may represent air or ground wave arrivals. Air waves could have traveled between the two antennas in the air, at the speed of light, from one face of the excavation to the other, over the top of the exposure (Fig. 22). In order to determine when a wave would arrive by air, the radar-wave travel times were calculated using the air distance around the top of the exposure for each test. These calculated arrivals are also plotted on the radar profile (Fig. 23). In the upper two tests, the first arrival and the projected air

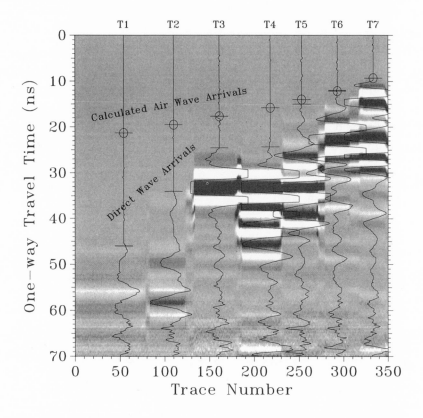

Figure 23. Wave arrivals from the transillumination tests. This is a gray scale GPR profile with overlain representative traces that can be used to measure the time of first arrivals in the seven steps shown in figure 22. The first direct wave arrivals are identified by the horizontal bar on each trace. Calculated potential air wave arrivals are shown by the circled horizontal bars. In tests 6 and 7, the air wave and direct wave arrivals are coincident and therefore no valid transillumination radar times were measured.

wave arrival were coincident, indicating they were air waves. Unfortunately, any later arrivals that might have traveled through the material were obscured by the high amplitude arrivals of the air wave. For this reason, the upper two tests are not accurate velocity readings for the material.

The identification of air waves that traveled between the two excavation faces illustrates the importance of using fairly deep excavations when conducting transillumination tests. If tests are conducted too close to the surface, radar energy can "leak" over the top or around the material to be tested, and the first arrivals in all tests attempted may be only air waves. It is also critical that accurate distances between antennas at all positions be obtained in the field so that air-wave calculations can be made and their arrival times calculated.

Knowing the horizontal separation of the antennas and the one-way travel time of the radar energy between the two antennas at each step, velocities could then be calculated (Table 4). The relative dielectric permittivity at each depth was also be calculated using equation 1.

Table 4

Results of Transillumination Tests

Test Number	Trace	Depth (cm)	Distance (cm)	Time (ns)	Velocity (cm/ns)	RDP
7	333	25	260	9.3	28.0	1.1
6	293	52	260	12.0	21.7	1.9
5	253	79	265	15.0	17.7	2.9
4	218	104	270	24.3	11.1	7.3
3	161	124	275	24.5	11.2	7.1
2	110	144	280	34.0	8.2	13.2
1	54	164	280	46.0	6.2	23.2

When the velocity measurements at each of the seven steps are plotted against the depth of the antennas, a velocity gradient graph can be constructed (Fig. 24). In this figure, at the maximum depth in the ground of 164 centimeters, the velocity was measured as 6.2 centimeters per nanosecond. It increased at a linear rate, with minor offsets, to a maximum of 17.1 centimeters per nanosecond 79 centimeters below the top of the exposure in test five.

Velocity data derived from transillumination tests is of great importance because it allows a velocity gradient to be measured as a function of depth, which is not usually possible in direct-wave methods. In the graph in figure 24, the velocity increases at a fairly constant rate with depth, probably indicating the gradually increasing residual water saturation that was visible as minor tonal changes in the exposed section. A minor change in the velocity gradient at 100 centimeters may indicate a change in velocity between volcanic ash Units 4 and 5, probably due to minor porosity and water-saturation differences in those layers.

If changes in velocity can be correlated with lithology or other changes in the material, they can yield important information when interpreting profiles. It is always important to understand the origins of reflections when attempting to understand GPR data from any site. Because all reflections are generated by velocity changes, transillumination tests are one of the best methods with which to understand these variations and possibly correlate reflections to known units.

Data from transillumination tests should, however, be used with caution because radar-wave travel paths within the material being tested can never be known for sure. Radar waves will tend to travel preferentially within the highest velocity material, and the time of the first arrival that is being used to calculate the velocity may be of the "fastest" material, not necessarily the material from the depth at which the antennas are placed. If the antennas are placed directly on an exposed layer that is "slower" than those bounding it, the radar waves may travel preferentially in the "faster" bounding layers along an indirect route. Any true arrivals that may have been received from waves that traveled through the lower-velocity layer would then be overwhelmed, obscured, or otherwise unrecognizable in the resulting traces.

Transillumination tests should always be performed in conjunction with direct-wave tests of objects at known depths. The combination of both types of velocity data will yield average vertical velocity measurements as well as a velocity gradient with depth. These data can be very important when constructing two-dimensional synthetic computer models and when correlating subsurface units visible in excavations to reflections in profiles.

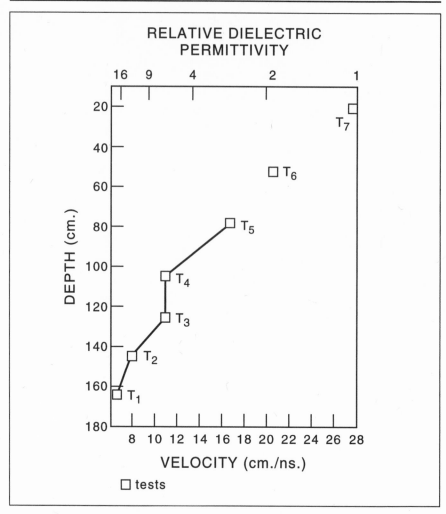

Figure 24. Velocity profile derived from the transillumination test data. The velocity decreases linearly with depth. The minor deflection at 100 cm (T_4) indicates a possible subsurface velocity discontinuity. The arrivals measured at T_6 and T_7 may represent the arrival of air waves, and therefore cannot be used to calculate the velocity of the material being tested.

Common Midpoint Tests

One of the simplest and fastest velocity tests that can be performed in the field is a common midpoint (CMP) test because no excavations are necessary. These tests are similar to transillumination tests in that radar energy is transmitted from one antenna to another, but in this case the antennas are located on the ground surface. CMP tests are performed by putting two antennas side-by-side on the ground and pulling them away from each other. A series of one way, or wide-angle two-way, radar travel times between the two antennas are then measured, and if the paths of energy travel in the ground can be deduced, the velocity of near surface layers can be measured. The antennas can also be placed apart and moved together, with their paths crossing at the common midpoint, and then moving away from each other (Fisher et al. 1994). A number of these types of separation tests can be performed at a site to produce a spatial distribution of near surface velocities within a grid.

There are a number of variations of CMP tests that can be performed, all of which are based on the same premise. One variation keeps one antenna stable while the other is pulled away. Data from this kind of test can be used in exactly the same manner as that derived from a standard CMP test; this type of test is commonly employed when only one person is available to move the antennas. Another test keeps the separation distance constant as both antennas are moved from station to station over the ground surface (Grasmueck 1994). If enough of these tests are conducted across a site, they will generate data that will allow for a map of changes in near-surface velocities over a GPR grid.

Common midpoint-type data are typically displayed in a standard GPR profile, with the antenna separation distance on the horizontal axis and time on the vertical axis (Fig. 25). As the antennas are moved apart, the first wave recorded is the air wave. Ideally it is recorded at time zero when the distance of antenna separation is zero. The second arrival is usually the ground wave that travels along the ground-air interface and is recorded soon after the air wave. The third, and any subsequent arrivals, are usually reflected or multiple reflected or refracted waves derived from subsurface interfaces. In areas with shallow water tables it may be possible to distinguish between what are referred to as "dry" ground waves and those that are "wet" and which may be traveling within saturated ground near the surface (Fisher et al. 1994).

In all cases, CMP-type tests usually can measure only velocities of surface soils or other material that is located very near the surface. They should not be viewed as a way of determining velocity at any great depth unless it is possible to identify actual wave travel paths.

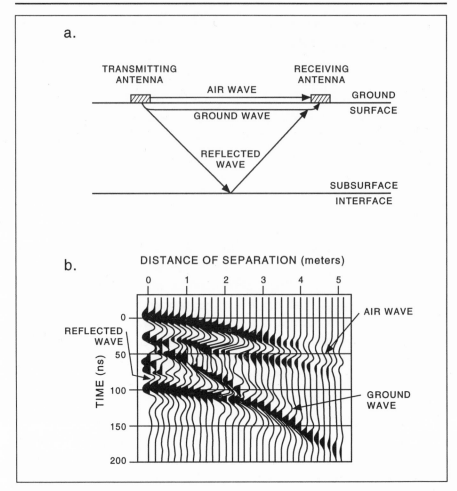

Figure 25. Common mid-point type tests. In (a) radar energy moves from the transmitting antenna along multiple paths (air, ground, and reflected waves) to the receiving antenna. When arrival times are plotted in profile (b) the airwave is recorded first, with the ground wave and all subsequent arrivals received later.

When conducting common midpoint-type tests, the most straightforward method is to place both antennas side-by-side on the ground. Radar pulses are then sent from the transmitting antenna to the receiving antenna, and the time window is adjusted so that the first arrival is recorded at time zero. This adjustment of the first arrival to time zero is not actually correct, because there will be a small amount of travel time between the two antennas depending on the separation distance of the two copper plates housed within their respective radar sleds; but this small time difference is usually not significant. As the antennas are separated, radar energy will continue to travel in the air between them, as well as along the surface of the ground and within near-surface layers (Fig. 25). The ground wave should intersect the air wave at time zero on the radar profile. Other radar waves that traveled deeper within the ground can be refracted within soil and stratigraphic units and sometimes reflected between subsurface layers before arriving at the receiving antenna, creating what can be a confusing series of reflections. These subsequent arrivals can be differentiated from ground or air waves because their arrivals do not intersect them.

In order to use these profiles to calculate the velocity of near-surface layers, it is necessary to know the maximum distance of antenna separation. This separation distance can be measured directly on the ground with a tape measure or tick marks can be placed periodically on the reflection record at known distances along the ground as the antennas are separated. If maximum separation distance is not measured directly, it can also be calculated using equation 1. As an example, a test performed at the Ceren site is shown in figure 26. In this test, the transmitting antenna was kept stable while the receiving antenna was moved away from it. The maximum distance of separation was not measured and the horizontal axis is therefore displayed only as the trace number of the recorded waves. At maximum separation (trace 300), the air wave was recorded at 13 nanoseconds. Knowing that the relative dielectric permittivity of air is approximately one, the distance can be calculated:

$$1 = \frac{.2998 \times 13}{\text{distance}}$$

The maximum separation distance is then calculated: 3.9 meters.

This same calculation can be performed for the second major arrival on the plot, which is the ground wave. Its arrival at maximum separation was received at 31 nanoseconds on the right hand side of the plot. Since the separation distance is known from the first calculation to be 3.9 meters, the relative dielectric permittivity can then be solved for using equation 1:

$$K = \frac{.2998 \times 31}{3.9}$$

$$K = 5.7$$

In the test shown in figure 26, a number of reflections were received after the second event that may be reflected or refracted waves that traveled deeper within the ground. The travel paths of these later arrivals and the distance the waves traveled are not known, and therefore velocity measurements of these deeper units are not possible. It is interesting that the RDP calculated for the ground wave is higher (and therefore the velocity is lower) than what was calculated for the bar test that was conducted nearby (Table 3). This may be because the ground wave was traveling almost exclusively within surface soils that contained a higher amount of organic material than the underlying volcanic units. Due to an increase in water held by organic material the surface soils would have retained more moisture, lowering their average velocity and raising the RDP.

LABORATORY MEASUREMENTS

At most archaeological sites, samples of subsurface units can usually be collected in the field and processed later in the laboratory to determine their relative dielectric permittivity, electrical conductivity, and magnetic permeability. These measurements can then be used to estimate velocity and signal attenuation for a site. They are also valuable when constructing two-dimensional computer models, discussed in chapter 5. If soil and sediment samples are collected and immediately stored in water-tight containers, they can be viewed as approximating field conditions. In reality, however, any collection and transportation of samples will modify the porosity, grain packing, and water saturation of the material somewhat, so natural field conditions can never be perfectly duplicated.

There are only a few devices that can make these types of laboratory measurements, none of which are commercially available. One way of determining the magnetic and electrical properties in the lab is by using tech-

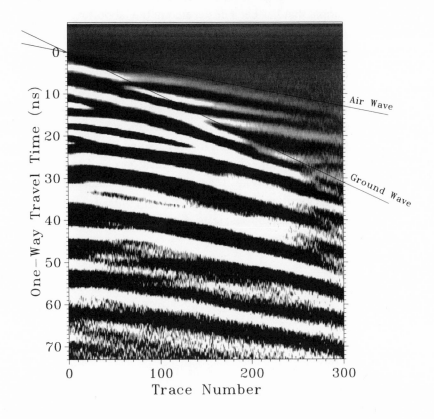

Figure 26. CMP-type test at the Ceren Site. The air wave and ground wave are recognizable as the first two distinct reflections, which intersect at time zero.

niques described by Kutrubes (1986) and Olhoeft and Capron (1993). In their tests, samples are first dried, crushed, and subjected to differing frequencies of electromagnetic energy. Measurements of RDP, conductivity, and magnetic permeability can then be made both when the samples are totally dry and at differing water saturations. Water-saturation changes can be simulated by progressively wetting the samples with distilled water from a dropper between tests.

There is a danger when using data from these types of laboratory tests because of changes in the material that affect grain packing and porosity during the testing procedure (Olhoeft 1986). Also, when water is artificially put back into a sample that has been desiccated in the laboratory, conditions are created unlike those in the field. Devices that measure the electromagnetic properties of samples also tend to put more energy into a sample than would usually occur in the field, where attenuation and wave dispersal are common.

In order to test the applicability of using laboratory tests to measure RDP, samples were collected from ten different volcanic ash units at the Ceren site (Conyers 1995a). These units varied from coarse-grained, air-fall pumice to finer-grained ash flows. The chemical composition of all ten units was known to be similar (Miller 1989). All samples were desiccated first and then differences in water saturation were simulated, using three drops of distilled water for partly saturated conditions; tests were also made after the samples had been completely saturated.

Table 5

Laboratory Measurements of Relative Dielectric Permittivities at Differing Water Saturations

		Relative Dielectric Permittivities	
	Dry	Partly Saturated	Totally Saturated
Unit 10	2.974	5.113	20.119
Unit 9	2.919	4.759	20.586
Unit 8	3.090	4.836	20.834
Unit 7	2.784	4.662	23.206
Unit 6	3.009	5.187	23.680
Unit 5	3.150	4.946	21.309
Unit 4	3.295	4.869	18.309
Unit 3	2.910	5.269	19.131
Unit 2	2.732	4.264	20.747
Unit 1	3.004	4.786	30.030
Mean	2.991	4.867	20.848
Standard Deviation	0.159	0.275	1.485

Results of these laboratory measurements indicate that all volcanic units have very similar relative dielectric permittivities when their water saturations are the same (Table 5). This is what was expected, because these volcanic units are known to be mineralogically similar, all having been derived from the same magma source (Miller 1989). Their RDPs as a group increase dramatically with water saturation, but with little difference noticeable between units. What differences there are was probably a function of minor differences in the amount of water that was dropped on the samples prior to testing.

The laboratory measurements indicate that the reflections visible in 300 and 500 MHz GPR profiles (Fig. 21) were probably generated at bed boundaries where a velocity discontinuity was created, probably due to minor water-saturation changes controlled by porosity differences. No significant bed boundaries, however, were visible as reflections in 80 MHz data because of the thinness of beds and the lack of resolution with this frequency antenna (e.g., Fig. 15).

An interesting approach to laboratory measurements was taken by Sternberg and McGill (1995) in Arizona. At their sites, samples taken of subsurface units were analyzed for particle size, mineralogic constituents, and water saturation. These sediments were then compared to published tables of the electrical and magnetic properties of similar geological materials (Olhoeft 1986). Relative dielectric permittivity and velocity were then estimated for their unique field conditions and radar travel times and were converted to approximate depth, using the methods described above. Sternberg and McGill thus could estimate the velocity without having to conduct laboratory measurements of their media's unique geophysical properties.

VELOCITY ANALYSES CONCLUSIONS

The most accurate velocity tests are those performed in the field that directly measure the radar travel times of objects at known depths. The object to be resolved should be metal, if possible, in order to maximize radar reflections. If possible, one or more test excavations should be made within or near a proposed GPR grid to the depth necessary to test the velocity of all materials that will potentially be studied. At depths greater than a few meters, iron bars or other relatively small objects may not be visible, and larger objects, such as the buried structure wall used as a target in the wall test, need to be imaged. When objects of this sort are not available, it may be possible to integrate shallow velocity measurements with stratigraphic correlation tests to arrive at deeper velocity values. Without these types of tests the correlation of important stratigraphic or cultural layers to reflections will always be suspect unless the stratigraphy is so simple and the archaeological features are so dramatic that the origin of resulting reflections is not in doubt. This is rarely the case.

If two or more excavations are available in close proximity, trans-illumination tests can be performed. The velocity data gathered from these types of tests can yield velocity-gradient curves that may delineate interfaces in the subsurface most likely to reflect radar energy. These types of data can be especially valuable in correlating reflections to known stratigraphy and other subsurface features.

If excavations are not available at a site, velocity of near-surface zones can be estimated using common midpoint-type tests. These tests can be used to estimate near-surface velocities, but they are usually not very valuable in obtaining velocity information from more than a few centimeters within the ground.

It is important to understand that any data derived from field velocity tests must be applied only to GPR reflection data that were acquired at about the same time. Ground conditions can change with the season or due to other factors, such as heavy rainfall or snowmelt, and subsurface radar velocity can change accordingly.

Lacking field velocity tests of any sort, samples of overburden material collected in the field can be analyzed in the laboratory or compared to standardized tables in order to obtain electrical and magnetic properties. These data can be used to arrive at an approximate relative dielectric permittivity, from which velocity can be calculated, if no other information is available. If samples are stored in water-tight containers soon after they are collected in the field, a more accurate estimate of normal field conditions can be made in the laboratory.

All or some of the above velocity tests should be performed as a matter of course during GPR surveys. For the most part, they are neither difficult nor time-consuming, and they will yield valuable time-depth information that is necessary in order to process raw GPR reflection data. Reflections that are recorded and then interpreted in time can only be used as a crude estimate of depth without accurate time-depth conversions. Without direct correlations to known stratigraphy or archaeological features using depth-corrected profiles, it can never be known for certain what is being mapped using GPR data.

Chapter 7

The Use of GPR Data to Map Buried Surfaces and Archaeological Features

In areas where well-defined GPR reflections are recorded in profiles, detailed subsurface mapping of the location of buried surfaces and archaeological features is possible. Prior to making maps, the genesis of the reflections must be known so that the maps being constructed are meaningful. This process, involving the correlation of reflections, can be accomplished using a variety of time-depth studies discussed in chapter 6 and with the aid of synthetic computer models discussed in chapter 5. Once the reflections in profiles are differentiated, they can be correlated—manually or with the aid of a computer—from profile to profile within a grid, and between grids in an area. If velocity and surface topography corrections have been made, accurate depth maps that define buried living spaces, or anthropogenic modifications to that landscape, can be made (Imai et al. 1987; Milligan and Atkin 1993).

ANCIENT LANDSCAPE RECONSTRUCTION

A standard manual GPR correlation and paleotopographic mapping technique was used to define the buried living surface at the Ceren site in El Salvador (Conyers 1995b). At this site, the ancient living surface, discussed in chapter 6, is now buried by between two and six meters of volcanic material, preserving the sixth-century landscape. The buried surface was initially defined in GPR profiles using both velocity analysis (Fig. 21) and synthetic computer modeling (Pl. 2a). It was then correlated within all GPR profiles, and its subsurface elevation was mapped.

Two vintages of GPR data were employed in this study. The first GPR survey at Ceren was conducted in 1979 in conjunction with a number of other experimental geophysical tests (Loker 1983). An initial grid, called Grid 1, was created that was approximately 100 meters by 100 meters, with a profile spacing every five meters (Fig. 27). This 80 MHz survey, done using an analog GPR unit, was one of the first large-scale GPR surveys conducted in archaeological exploration (Sheets et al. 1985). Additional GPR grids, made using 300 MHz antennas and a digital system (Grids 3, 4, and 5), were acquired in 1994; these overlapped and extended Grid 1. The older-vintage 80 MHz analog reflection data, which were saved on magnetic tape, were digitized and computer processed prior to interpretation. A total of 7600 meters of GPR data were interpreted within the four grids.

DATA INTERPRETATION

The top of the buried living surface (the TBJ ash) is the most important geologic interface to map with GPR at this site because it was the ancient living surface prior to burial. During radar profile interpretation, the TBJ reflector, first identified along the stratigraphic test line discussed in chapter 6 (Fig. 22), was hand-colored on each section; and its elevation in meters above sea level was compiled every 1.25 meters. This reflector was visually correlated line-by-line throughout all GPR grids and directly compared at each intersecting line to assure consistency. Subsurface elevations along each profile were then plotted on the grid map and contoured both by hand and with the aid of a computer to reveal the three-dimensional topography of the buried TBJ living surface (Fig. 28).

The high-amplitude nature of the TBJ reflector is quite consistent within and between grids, as the synthetic models discussed in Chapter 5 predicted (Pl. 2a). Many times, however, the TBJ reflector exhibited amplitude variations that were possibly related to anthropogenic modifications on the TBJ surface, attenuation with depth, or changes in focusing due to bed geometry (e.g., Fig. 21).

In some of the 80 MHz GPR profiles in Grid 1, the TBJ reflector occurred within 2.5 meters or less of the ground surface. Above 2.5 meters, the TBJ reflector was partially or wholly obscured within the near-field zone

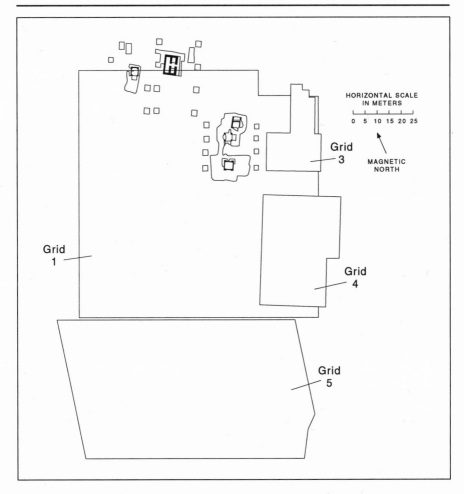

Figure 27. Location map of the GPR grids at the Ceren Site, El Salvador. GPR data in Grid 1 were acquired in 1979 using a single 80 MHz antenna. Data from Grids 3, 4, and 5 were acquired in 1994 using dual 300 MHz antennas. Archaeological excavations and the structures exposed within them, are indicated.

(Figure 13). Subsurface elevation data could not be compiled in these areas, and the portions of those lines where this occurred were noted only to be "very high." This was not a problem in the grids acquired using the 300 MHz antenna that had a shallow near-field zone; the TBJ reflection in these was visible and easily correlative in all lines (e.g., Fig. 18).

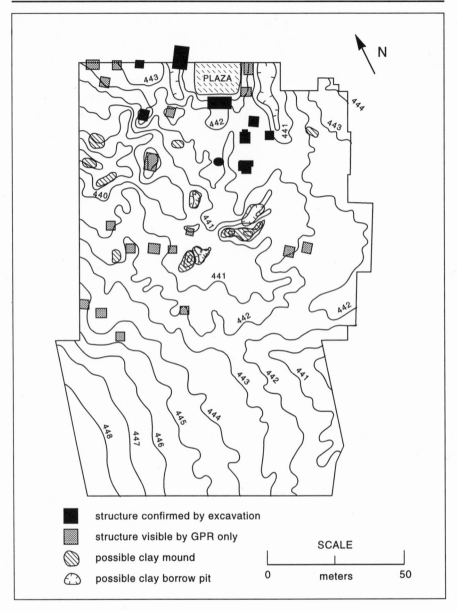

Figure 28. Paleotopographic contour map of the buried TBJ living surface, Ceren Site, El Salvador. The grid locations are the same as in figure 27. Contours are in meters above sea level, with a contour interval of one meter.

BURIED STRUCTURE IDENTIFICATION

All point-source reflections, which may denote buried structures or other sub-surface anomalies, were recorded on the base map during profile interpretation. As demonstrated in the synthetic model (Pl. 2a), point-source reflections were probably derived from the top of structural platforms and from walls or columns. When paired point-source reflections were visible, the intersection of the two reflections denoted the possible location of the floor of the buried structure.

Due to the lack of subsurface resolution in the 80 MHz data, it was many times difficult to differentiate the TBJ reflection from horizontal reflections derived from the tops of structure floors. Where not obscured within the near field zone, possible clay floor reflections were many times recognizable only by an increase in the amplitude of the TBJ reflection. This amplitude increase was caused by the large velocity contrast between the hard-packed clay surface of the structure floor and the overlying volcanic material, as predicted in the synthetic model (Pl. 2a). The abrupt decrease in radar energy velocity at the interface yielded a high coefficient of reflectivity, producing the high-amplitude reflections.

The 300 MHz reflection data, with its greater resolution, were capable of mapping subtle changes in elevation of the TBJ surface of as little as 20 to 25 centimeters. Increases in amplitude of reflections derived from houses' clay floors were also visible (Fig. 18) similar to the floor reflections in the 80 MHz data.

The stratigraphy of the overlying volcanic units is readily visible in 300 MHz profiles (Fig. 18), but it is totally invisible in the 80 MHz profiles due to a lack of thin-bed resolution with this low-frequency antenna. As demonstrated in the stratigraphic tests (Fig. 21), low-amplitude reflections occurred at bed boundaries where minor changes in water saturation caused minor velocity changes that produced reflections. Volcanic beds visible in profiles exhibit a pronounced drape over the underlying structures, a phenomenon routinely documented during archaeological excavations (Miller 1989). This draping is accentuated by standing walls or columns that produced dunes of ash around them (e.g., Fig. 20).

Distinctive point-source reflections with apexes above the clay floors of structures occur on the 300 MHz data (Fig. 18). The apexes of these point-source reflections are a little more than one meter above the clay floor surface, similar to the height of many standing columns in some excavated structures (Kievit 1994). It is probable that these point-source reflections are the record of reflections that occurred from the tops or sides of standing columns or walls, as predicted in the two-dimensional synthetic models (Pl. 2a). Point-source reflections from clay floors also occur, but are less common in 300 MHz data than in 80 MHz data. They were not recorded farther than about one meter away from the clay foundation from which they were generated. This is because the radar beam of the 300 MHz antenna is narrower (Fig. 6) and is thus not capable of "seeing" the anomaly as far in advance or behind the antenna surface location as in 80 MHz data, creating a less distinct point source hyperbola.

PALEOTOPOGRAPHIC MAPS

The mapping of the TBJ living surface and the location of structures that were built upon it was accomplished by the construction of paleotopographic contour maps of the buried landscape and the identification of radar anomalies that were generated from buried structures. In order to construct maps of the buried topography, more than 3000 data points, which consisted of values representing subsurface elevations of the TBJ surface (or what appeared to be the TBJ but may have been the tops of buried-structure floors), were plotted on the base map shown in figure 27. These subsurface data were then hand-contoured with a contour interval of one meter. The resulting map represents the topography of the TBJ living surface prior to the eruption of the nearby volcano (figure 28). All structure locations, as defined during archaeological excavations or as identified by point-source hyperbolas in GPR reflection profiles, were also plotted within the mapped area (Fig. 29).

ANCIENT DRAINAGE PATTERNS AND TOPOGRAPHY

The most striking features of the buried TBJ surface are the variation in topography across the site and the intricacy of the buried drainage pattern. The highest elevation of the TBJ surface, in the southwestern portion of the sur-

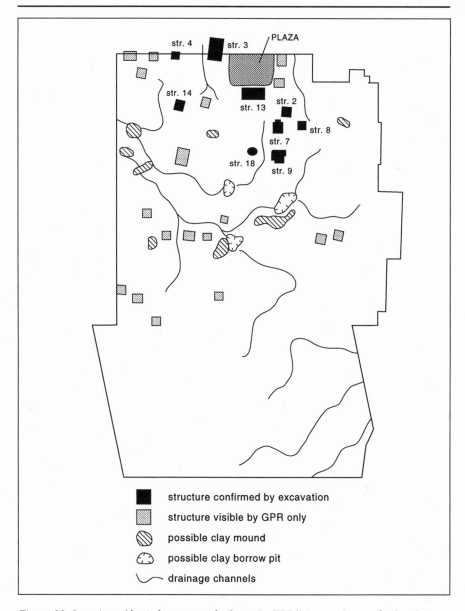

Figure 29. Location of buried structures built on the TBJ living surface and other features visible on GPR profiles of the Ceren Site, El Salvador. The structures were identified by reflection hyperbolas derived from point-sources and amplitude changes in the TBJ reflection where structures are located, as modeled in the two-dimensional synthetic model shown in plate 2A.

veyed area (Fig. 27), was measured at 448.45 meters above sea level. The lowest elevation, where the TBJ could be directly measured using GPR, is in the western portion of the survey area, where it is located at an elevation of 437.7 meters. The total range of elevation within the mapped area is 10.75 meters.

In general, the landscape just prior to the volcanic eruption consisted of a small elongated valley, located in the west-central portion of Grid 1, surrounded by low bluffs to the north and east. A gradual southern slope rose upward out of the valley to the southeast, ultimately forming a large hill to the south.

Buried structures are located primarily on the northern bluff and southern slope, but a few others are located on subtle topographic rises on the edge of the central valley. No buried structures have yet been discovered in the southernmost portion of the survey area.

All nine of the buildings partially or totally excavated to date were located on the northern bluff. The majority of the drainage channels mapped within the GPR grids flowed from east to west. Interestingly, the present-day surface drainage in this same area flows in the opposite direction, probably due the deposition of the thick wedge of volcanic material to the north and west of the site during the eruption that buried the site. All surface water-flow directions were then reversed. The 300 MHz GPR data are capable of resolving buried channels about 50 centimeters or less deep, while only the largest drainage features are visible on the 80 MHz profiles. Drainage channels visible on 80 MHz profiles had a maximum depth of about 1.5 meters and in some areas were 2 meters or more wide. Their banks were gently sloping, with no discernable cutbanks. Numerous small, closed depressions that are surrounded by small mounds within the central valley may have been caused by clay quarrying operations (Fig. 29). The small mounds near them may be piles of clay that had been excavated, but not yet transported to construction sites. All of the buildings excavated at the site were constructed of clay, and large amounts of clay were also commonly used as a surfacing material for patios and plazas (Kievit 1994; Sheets 1992).

COMPUTER-GENERATED THREE-DIMENSIONAL MAPS

A total of twenty-five buried structures are identifiable using both the 80 MHz and 300 MHz GPR data. Nine of the GPR-defined structures have been confirmed by excavation. Only one structure is known from archaeological excavations but not visible on GPR profiles (Structure 9). It is a sweat bath with a partially collapsed dome-shaped roof. The roof top extends more than a meter and a half into the overlying volcanic material, placing it within the near-field zone for 80 MHz data, which may have obscured it on profiles. In addition, the roof's dome shape likely dispersed radar energy (e.g., Fig. 12) making any radar reflection from it even less distinct. This failure of radar to image a known prominent structure is one example of how the geometry of buried features and their depth of burial can make them all but invisible in a GPR survey.

On the southern slope of the mapped valley eleven additional buried structures are visible in GPR profiles (Fig. 29). All are located on topographic rises and are separated from each other by numerous channels that flowed to the north into the central valley.

In order to produce visual images of the ancient landscape, computer-generated three-dimensional pictures of the TBJ surface were constructed using the subsurface data derived from profiles within all the grids. Three values were input in the computer (x, y, and z) for each data point. The x and y data were the surface grid locations. The z value was the elevation of the TBJ reflector in meters above sea level. All x, y, and z values were transferred to a three-dimensional mapping program. A three-dimensional image was constructed for the site illustrating all the geographic features visible on the paleotopographic map (Pl. 2b). The program used to create these maps applied a "minimum curvature" arithmetic algorithm that created a "best fit" surface to the data points.

Once the buried topography had been mapped and the location of the structures plotted, a digital three-dimensional map was produced of the buried topography and the structures built on it. A series of images was then constructed that illustrated the ancient village as it was just prior to burial about A.D. 590 (Fig. 30). Using computer animation techniques, the land-

Figure 30. Three-dimensional representation of a portion of the Ceren Site GPR grids. The structures and trees that have been excavated or are visible on GPR profiles have been digitally modeled as stick figures. The view is from the northeast, 35 degrees from the horizon. There is no vertical exaggeration of the buried topography. (Image courtesy of Fenton-Kerr Engineering.)

scape, structures, and trees were rendered, using different colors and textures to give a more realistic view of the ancient village mapped by GPR (Pl. 3a).

Integration of the GPR-produced maps and the archaeological data from excavations demonstrates that the population density at Ceren was quite high at the time of the eruption. The presence of a large, central plaza and its associated communal buildings also indicate the presence of many more people than conceivably could have lived in the households so far identified in excavations and with GPR.

The extensive GPR coverage of the Ceren site has identified many possible land use areas and structures that are scheduled for excavation in future years. Due to their deep burial the discovery of these archaeological features would have been impossible without GPR exploration. Just as important is the ability of GPR mapping to place known and newly discovered archaeological features within their paleogeographic context. Because the total extent of the site may never be excavated due to both cost and preservation concerns, much of what we will learn about the environment surrounding the excavated structures must come only from the GPR-produced maps of the buried topography.

Chapter 8

Amplitude Analysis in GPR Studies

The primary goal of most GPR surveys in archaeology is to identify the size, shape, depth, and location of buried cultural remains and related stratigraphy. The most straightforward way to accomplish this is by identifying and correlating important reflections within two-dimensional reflection profiles. These reflections can then be correlated from profile to profile throughout a grid, as described in chapter 7. Another more sophisticated type of GPR data manipulation is amplitude slice-map analysis, which creates maps of reflected wave amplitude differences within a grid. The result can be a series of maps that illustrate the three-dimensional location of reflection anomalies derived from a computer analysis of many two-dimensional profiles. This method of data processing can only be accomplished using GPR data that are stored digitally on a computer.

The raw reflection data acquired by GPR is nothing more than a collection of many individual traces along two-dimensional transects within a grid. Each of those reflection traces contains a series of waves that vary in amplitude depending on the amount and intensity of energy reflection that occurred at buried interfaces. When these traces are plotted sequentially in standard two-dimensional profiles, the specific amplitudes within individual traces that contain important reflection information are usually difficult to visualize and interpret. The standard interpretation of GPR data, which consists of viewing each profile and then mapping important reflections and other anomalies, may be sufficient when the archaeology and geology are simple and interpretation is straightforward. In areas where the stratigraphy is complex and buried features are difficult to discern, different processing and interpretation methods, one of which is amplitude analysis, must be used. In the past, when GPR reflection data were collected that had no discernable reflections or recognizable anomalies of any sort, the survey was usually declared a failure and little, if any, interpretation was conducted. Recently, with the advent of more powerful computers and sophisticated software programs that can manipulate large sets of digital data, important subsurface information has been extracted from these types of GPR data in the form of amplitude changes within the reflected waves. In this chapter, examples will be shown where digital reflected-wave information was utilized for purposes beyond mapping the location of anomalies and the elevations of subsurface layers.

An analysis of the spatial distribution of the amplitudes of reflected waves is important because these changes are the direct result of changes in the makeup of subsurface units. The higher the contrasting velocity at a buried interface, the greater the amplitude of the reflected wave. If amplitude changes can be related to important archaeological features and stratigraphy, the location of higher or lower amplitudes at specific depths can be used to reconstruct the subsurface in three-dimensions. Areas of low-amplitude waves indicate uniform matrix material or soils, while those of high amplitude denote areas of high subsurface contrast such as buried archaeological features, voids, or important stratigraphic changes. In order to be interpreted correctly, amplitude differences must be analyzed in "time-slices" that examine only changes within specific depths in the ground. Each time slice is comparable to a standard archaeological excavation level, except using GPR data the levels consist of a collection of reflected wave amplitudes instead of sediment, soil, and artifacts.

It is important to understand that the archaeologist should not simply send raw GPR data to a computer-processing expert and have that person blindly make time-slice maps. These types of maps are not made automatically, like having photographs developed, but must be constructed thoughtfully. The archaeologist should not be a passive partner in this powerful data-processing step, but must actively participate in determining the processing parameters and visual format of the final output. Their production requires some prior knowledge of site conditions and the types and dimensions of the features to be resolved.

An example of an amplitude anomaly map that was constructed from a horizontal slice in the ground between two and three meters at the Ceren site in El Salvador is shown in plate 3b. In this map, two structures are visible as high-amplitude anomalies. These structures were also noticeable, and could be mapped, using the standard interpretation methods employing two-dimensional profiles. Other high-amplitude anomalies, however, are visible on this amplitude map that likely represent additional archaeological features not readily visible on individual profiles. This ability to identify and map amplitude anomalies that are difficult to discern visually in individual profiles is one of the most powerful applications of this type of GPR analysis. In 1997, this area of the grid was resurveyed using a higher amplitude 500 MHz antenna with a 50 centimeter line spacing. A more critical analysis of the profiles in this new grid indicates that the large high-amplitude anomaly in the southern portion of the grid may be a large, open plaza or patio. It is likely that, without its identification in the slice maps, this area of subtle amplitude change would have gone unrecognized and uninterpreted.

It also can be very informative to compare on a single map the location of amplitude anomalies from many horizontal or sub-horizontal slices in the ground. In this way the orientation, thickness, and relative amplitudes of anomalies are visible in three-dimensions. Amplitude slices are usually made in equal time intervals, with each slice representing an approximate thickness of buried material. Viewing amplitude changes in a series of horizontal time slices within the ground is analogous to studying geological and archaeological changes in equal depth layers (Arnold et al. 1997; Goodman et al. 1995; Malagodi et al. 1996; Milligan and Atkin 1993). If velocity analyses are performed in advance and time depth corrections are made, each horizontal time-slice can be viewed as an approximate depth slice. If the amplitude anomalies in each depth slice are then correlated to known archaeology and stratigraphy from nearby excavations, extremely accurate three-dimensional maps of a site, broken down into levels, can be constructed.

Amplitude anomaly maps need not be constructed horizontally or even in equal time intervals. They can vary in thickness and orientation, depending on the archaeological and geological questions being asked. Surface topography and the subsurface orientation of the features and stratigraphy of a site may sometimes necessitate the construction of slices that are neither uniform in thickness nor horizontal. This can easily be done on the computer when the data are in a digital format.

Figure 31 illustrates a schematic diagram showing the general principle in the creation of a horizontal time-slice. Radar profiles collected along parallel and/or cross lines within a grid are sliced at a particular time interval (d_t). The relative amplitudes of the reflected radar waves that were recorded between those times (the slice) are then averaged and interpolated prior to printing them in map form. The resulting anomalies visible in a slice map therefore represent the spatial distribution of reflection amplitudes between specific depths across the grid. In standard horizontal time-slice maps, each slice is distinguished by a two-way time interval, measured in nanoseconds. Due to possible velocity differences across a grid, however, it is likely that what appears to be a perfect horizontal slice in a computer-generated map may actually analyze collected data that are neither horizontal nor of equal thickness. Due to velocity changes across the area and with depth, a slice map may actually represent information that varies considerably in elevation and which thickens and thins across a grid. Horizontal time slices must therefore be considered only approximate depth slices. Without very detailed velocity control throughout a grid, it is impossible to construct perfectly even and horizontal slices that can be measured in true depth. In addition, if GPR data are collected along an uneven ground surface and no topographic corrections are made, each slice in the ground will vary considerably from the horizontal and be parallel to the ground surface instead of the horizontal plane.

To compute horizontal time slices, the computer must compare amplitude variations within traces that were recorded within a defined time window. When this is done, both positive and negative amplitudes of reflections are compared to the norm of all amplitudes within that window. No differentiation is made between positive or negative amplitudes in these analyses, only the magnitude of amplitude deviation from the norm. Low-amplitude variations within any one slice denote little subsurface reflection, and therefore indicate the presence of fairly homogeneous material. High amplitudes indicate significant subsurface discontinuities and in many cases detect the presence of buried features. An abrupt change between an area of low and high amplitude can be very significant and may indicate the presence of a major buried interface between two media.

Degrees of amplitude variation in each time slice can be assigned arbitrary colors or shades of gray along a nominal scale. It is usually not

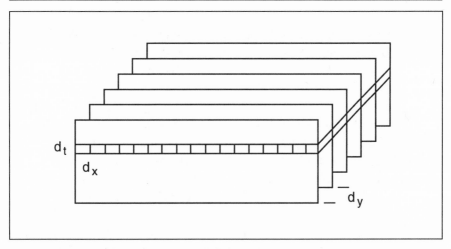

Figure 31. Diagrammatic representation of the construction of horizontal slice maps from standard two-dimensional GPR profiles within a grid delineated by x and y values. The distance d_y is the distance between profiles; d_x is the distance along the profile in which reflected waves are averaged; d_t is the thickness of the slice, measured in nanoseconds (two-way travel time). The average of many squared amplitudes in the windows d_x, d_t located at locations x, y within the grid, is the amplitude parameter displayed in amplitude anomaly maps.

important what color or shading scheme is used in slice maps, as long as there is enough contrast within the map to make amplitude anomalies recognizable spatially. A high-to-low amplitude scale is normally presented as part of the legend of each map, but without specific units because, in GPR, reflected wave amplitudes are usually arbitrary.

It must be remembered that because most subsurface layers are not perfectly horizontal, and most stratigraphic units vary in thickness, horizontal time slices may not be comparing the amplitudes of units that are correlative. If this is the case, abrupt changes in color or tone in the resulting maps may represent only the intersection of a time slice with a stratigraphic boundary and may not be indicating a meaningful geological or archaeological change. These complications can sometimes be adjusted for if the orientation and thickness of the subsurface layers are known.

There are many options that can be employed when creating time-slice maps, but in general there are some useful rules that must be followed when interpreting most digital GPR data sets:

- The spacing between lines (d_y in figure 31) in a grid should be at most one half of the wavelength of possible reflections that are created from the smallest targets that one wants to image. This distance can be estimated using equation 3 in figure 4. The smaller the target, the closer the spacing should be between lines.
- The sampling rate should be adjusted so that it is sufficient to record enough digital data within each trace to define the reflected waves generated at the desired target.
- After digital reflection data are collected, and prior to creating the amplitude slice maps, it is important to do spatial averaging along each profile with the computer. The averaging should be such that the interval between profiles (specified as d_y in figure 31) should be about the same as a spatial averaging window along the profile (specified as d_x in figure 31).
- Digital amplitude reflection data can also be averaged in the vertical (time) window. It is useful in many archaeological data sets to average vertical reflections over several wavelengths (d_t in figure 31). Slicing the data at a very narrow time window (d_t) can many times create a very"noisy" time slice. A thin slice may have a high resolution, but the noise may overwhelm any real amplitude data that are usable. In addition, very thin slices have a greater chance of crossing stratigraphic boundaries, creating artificial amplitude anomalies at the intersection of slices and subsurface beds.

To create visually useful maps where profiles are fairly far apart necessitates two-dimensional interpolation between the profiles within a grid. The amount of interpolation between profiles dictates the resolution of the resulting anomalies when plotted in map form. In figure 32, amplitude slice maps with three different interpolations are displayed, all of which are derived from the same digital data set within a 13- by 33-meter grid collected with a 50 centimeter spacing between transects. The resulting amplitude anomalies, plotted in a slice between 7.5 and 10 nanoseconds (two-way time), are shown on the maps. These data were acquired using 500 MHz antennas at the fifteenth-century Nanaojo Castle ground in Shimane Prefecture in Japan (Fig.

33). The area surveyed was believed to include the buried entrance to Nanaojo Castle. It was hoped that GPR would be capable of imaging stones in the bottom of post holes used to contain vertical support timbers at the castle entrance. The buried right-angle anomaly at the top of each slice, especially in "A" and "B" (Fig. 32) is the corner of one of these support structures.

To illustrate the different resolutions produced by differing interpolation radii, three slices were made from the same data set from Nanaojo Castle. When the search radius was small, individual stones used as foundations for the vertical support timbers can be seen ("A" in Fig. 32). As more digital data points

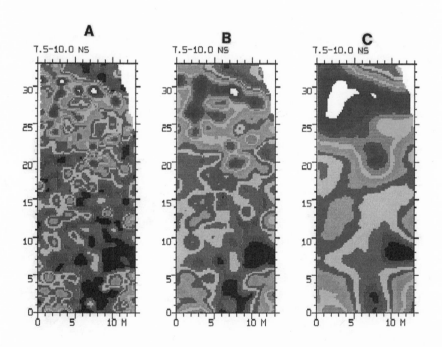

Figure 32. Amplitude slice maps with differing interpolations between profiles. The line spacing within the grid is 50 centimeters. Anomalies in Map A were generated with a search radius of 60 centimeters during interpolation. Map B shows the anomalies generated with a 1.1 meter search radius. Map C was constructed with a 2.5 meter search radius. The GPR data was collected near the entrance of Nanaojo Castle, Japan. The anomalies represent a stone foundation that may have been part of the castle entrance. All three maps are of the same time-slice in the ground.

were averaged, these individual amplitude anomalies generated by the computer became smoothed and only the right-angle feature, without the individual stones, is visible ("B" in Fig. 32). After the search radius was increased again, the data was smoothed even more; and the right-angle feature became visually distorted and unrecognizable as a right-angle form. In this example the smaller search radius is preferable because the important features that need to be resolved (the individual stones making up the feature) are also small. The reverse would be true when small objects such as river cobbles make up the matrix of a site. In this case, if the target features were large and surrounded by river cobbles, a large search radius would be needed when interpolating between lines to average out the individual reflections from the cobbles (the clutter) and focus on the large features.

During the preparation of geophysical data displays such as amplitude time-slices, it is seldom the case that one single interpolation and gridding parameter will be sufficient. Digital reflection data from a GPR grid normally have to be computer processed using a number of interpolation and gridding parameters in order to obtain results that are acceptable for the specific conditions of the site.

AMPLITUDE SLICE MAPS ON LEVEL GROUND

The Nyutabaru Burial Mound site in Miyazaki Prefecture on the Japanese island of Kyushu (Fig. 33) is a good example of the utility of amplitude time slices in defining buried sites that are completely invisible by any other means. Many burial mounds and other ceremonial structures are found in this area (Arita 1994), some containing intact burials from the Kofun Period (A.D. 300–700). Today, much of the area has been converted to farmland, and many of the ancient burial mounds have been leveled. Although in some areas there is no remaining surface evidence of these features, burial chambers are sometimes still intact in fields below the plow zone. Many times they are accidentally discovered when heavy farm equipment falls into intact chambers during plowing or harvesting operations. Prior to leveling, some of the larger mounds were eight to nine meters in height and over 100 meters in length (Hiroshi 1989).

This area of Japan is volcanically active, and a number of significant historic and prehistoric ash falls have helped to bury and obscure archaeologi-

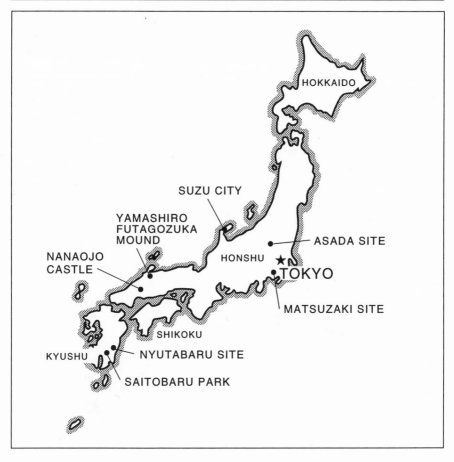

Figure 33. Location of archaeological sites in Japan where the GPR data discussed in the text were acquired.

cal features (Machida and Arai 1983). The volcanic ash units that make up the matrix of the site have been partially weathered and surface soils homogenized by farming operations over the years. Although there are well-developed surface soils, the underlying sediment is for the most part clay-poor, making it an excellent medium for radar-wave transmission.

At the Nyutabaru site, a GPR grid 108 by 45 meters was surveyed in 1993. The area, an almost perfectly flat agricultural field, was planted in alfalfa at the time of the survey (Fig. 34). The surface soil was quite moist at the time data were acquired. Profiles were collected using a single 300 MHz antenna in east-west lines every meter within the grid. Velocity estimates were obtained in a test trench in the same fashion as described for the "bar test" in chapter 6. An average radar-wave velocity of 8 centimeters per nanosecond was obtained.

A representative GPR profile across the grid is shown in figure 35. In this profile, a moat that at one time surrounded a central mound is visible twice where the transect crossed it. The moat exhibits the classic bow-tie effect, illustrated in the synthetic radargram in plate 1a. Horizontal amplitude time-slice maps were constructed every 8 nanoseconds (two-way time) from the data recorded within the grid (Pl. 4). Using an average velocity of about 8 centimeters per nanosecond, each horizontal slice displays the amplitude anomalies in successively deeper slices, each about 32 centimeters in thickness. The uppermost map displays linear features that are present in the surface soil, oriented with the present-day furrows visible in the alfalfa field. The next-deepest slice, from 8 to 16 nanoseconds, which is below the modern plow zone, depicts amplitude anomalies that are orientated perpendicular to those in the overlying

Figure 34. Acquiring 300 MHz data at the Nyutabaru Site, Japan. The field over which the survey was conducted was planted in alfalfa. A preserved burial mound, which was not part of the GPR survey, is visible in the background.

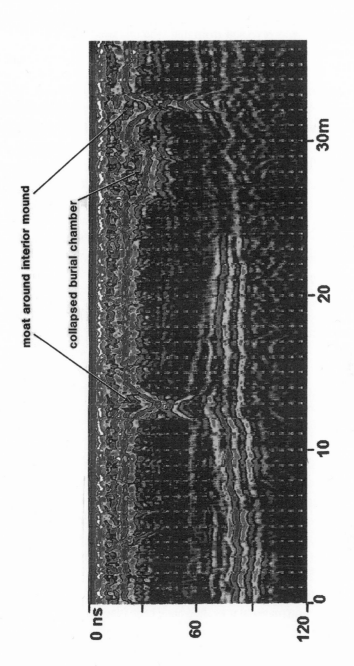

Figure 35. Two-dimensional profile across a moat and a possible burial chamber at the Nyutabaru Site, Japan. The circular moat was crossed twice by this profile and is visible by the "bow tie" effect that is modeled in plates 1a and 1b. A central collapsed burial chamber is visible, but poorly defined in this profile. (After Goodman et al. 1995.)

slice. This slice approximates a depth of between 32 and 64 centimeters in the ground. Aerial photographs that were taken at the site in the 1940s show that the orientation of row crops was reversed from that which is used today, possibly accounting for the change in trend of near-surface soil structures in this deeper slice.

In the slice from 16 to 24 nanoseconds some of the pre-1940s plow features are still visible, but little else of interest. Some interesting amplitude features begin to show up in the next-deepest slice, from 24 to 32 nanoseconds, which displays amplitude anomalies between about 1 and 1.3 meters in the ground. In this slice, a circular feature about 22 meters in diameter is visible as low-amplitude reflections in the eastern portion of the grid. In progressively deeper slices, this feature becomes more pronounced, and the reflections that generated it become higher in amplitude than the surrounding material. This circular amplitude anomaly is believed to have been generated from the walls and perhaps the floor of a burial moat that surrounded a now-flattened and destroyed central mound.

In the 40 to 48 nanosecond slice, which represents a depth-slice of between about 1.6 to 2 meters in the ground, a very high amplitude anomaly is visible just to the south of the northern edge of the moat. This feature is believed to be a burial chamber that was at one time covered by a large mound. The weaker anomaly outside of the moat, just north of the chamber, may represent the remains of an entry shaft that once led to the chamber under the moat. A portion of a second moat, which may surround another burial chamber, is located to the northeast in this grid. It has not yet been fully surveyed by GPR.

In this deepest slice, from 40 to 48 nanoseconds, the two linear features located to the west of the moat may be corral enclosures. Written records from the Edo Period in Japan (1600–1860) indicate that an ancient corral for keeping horses existed somewhere in this area. The high-amplitude nature of these two linear anomalies, which are strikingly similar to those of the moat, indicate that the moat and the corral were probably constructed in a similar fashion. It is likely that the Edo Period corral was constructed by digging two parallel trenches and piling the excavated dirt between them. A wooden fence or some other construction material may then have been placed on the mounded dirt. The corral fence lines are presented in the amplitude anomaly map as two parallel anomalies generated from reflections that were created along the bottoms of two trenches that border the raised fence. The fence itself likely deteriorated long ago.

The ability to map this site in three-dimensions using the amplitude slice-maps allows an interpretation of a scenario for the construction, abandonment, burial, and partial destruction of the mounds and their associated features. This sequence of events is illustrated in figure 36. First the moats were dug into level ground (A), and the material that was excavated from them was piled in the middle to create a mound. After the mound was built, a vertical shaft, which originated outside the moat, was dug to several meters depth. A tunnel was then constructed horizontally, from the end of the vertical shaft into the mound where a central chamber was constructed (B). After a body had been interred in the central chamber, the outside entrance to the shaft and tunnel was filled (C). Over time, the moats were filled by natural sedimentation and the deposition of volcanic ash from the nearby Kyushu volcanoes. The difference in the soil characteristics between the moat fill-material and the surrounding soil likely produced the high-amplitude reflections visible as anomalies in the 40 to 48 nanosecond slice in plate 4. Intensive farming practices that were instigated in the nineteenth and twentieth centuries leveled the ground, destroying the central mound but preserving the underlying moat, central chamber, and its entrance (D). The addition of volcanic ash from historic eruptions and intensive plowing further obscured the underlying archaeological features (E).

The results of the Nyutabaru survey, when depicted as a series of amplitude slice maps, are in many ways quite remarkable. Although many of the larger archaeological features, such as the moat and possibly the central burial chamber, are visible in some of the standard two-dimensional profiles (Fig. 35), their orientation cannot be easily discerned. The smaller, less-distinct features, like the central burial chamber and the corral enclosure, are almost completely invisible. When the digital amplitude data are viewed in amplitude slice maps, however, the features become immediately visible. Because the data are illustrated as a sequence of maps descending into the ground, the three-dimensional aspects of the site can be used to reveal both exact orientations and the building and burial sequence.

With GPR slice maps, the exact location, both spatially and in depth, of the buried architectural features at this site are now known with precision. The area of the GPR grid, which would normally be much too large to excavate, was surveyed in only four hours using a zig-zag type grid pattern. The central burial

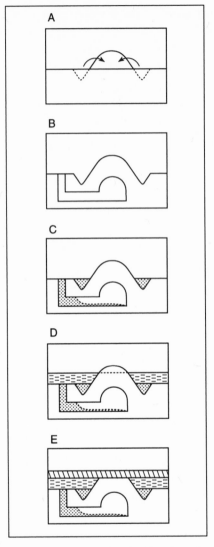

Figure 36. The building and burial sequence of the Nyutabaru burial mound and moat that was constructed from the three-dimensional slice-map data. Cross section (A) shows the initial construction of the moat and the interior mound. Soil and sediment from the construction of the moats were used in building the central mound. In (B), a vertical shaft and horizontal tunnel were constructed into the central mound, and a burial chamber was excavated. The moats were ultimately filled (C), and the vertical shaft collapsed and was filled with soil and sediment. In (D) the area was covered with volcanic ash, and the top of the interior mound was partially removed by erosion and later plowing. Further volcanic ash deposition and soil formation (E), removed any surface evidence of the buried archaeological features.

chamber has not been excavated, but knowing its exact location and depth will allow archaeologists to take measures to insure its preservation.

AMPLITUDE TIME SLICES ON UNEVEN GROUND

When GPR data are collected in a grid that is situated on uneven ground, topographic corrections must usually be made. When only reconnaissance exploration is the goal, however, these corrections may not be mandatory prior to constructing amplitude slice maps as long as the interpreter understands what is being viewed in each slice.

A good example of this type of reconnaissance survey is the Saitobaru Kofun Mound 111. It is located in Miyazaki Prefecture on the southern island of Kyushu, an area that contains one of the most famous collections of burial mounds in Japan (Fig. 33). According to Hongo Hiromichi, head archaeologist of Miyazaki Prefecture, the Saitobaru Plateau contains more than 311 mounds dating between the fourth and sixth centuries, most of which overlook a large valley. The Saitobaru area was designated a National Historic Park in 1952. Several burial mounds have been excavated at the site, but the majority remain undisturbed. The park is unusual in Japan because it contains many undisturbed, and presumably intact, burials; only a few mounds have been destroyed by farming or construction activity. The mounds are thought to have been situated so that the monumental funeral architecture of the society's elite ancestors could be visible to those who lived and worked in the nearby lowlands.

A variety of mound styles and shapes exist in the Saitobaru area. Some of the mounds are keyhole-shaped in plan view, while others are circular or square. Each mound presumably contains within its interior one or more stone-lined burials, shafts, or other features of archaeological interest. Prior to the construction of the mounds, beginning in about the fourth century, the area was periodically buried by volcanic ash. This volcanic material, which has been partially weathered, was the material used in the construction of the mounds and other archaeological features in the area. Preliminary analysis indicates that this material is low in clay and allows good radar penetration with little attenuation, even when soil and sediment units are partially water saturated.

Within the last few years, some of the burial mounds in the Saitobaru Park have been surveyed using GPR in an attempt to delineate internal features without having to excavate. One of these mounds, Kofun 111, is visible on the surface as a large internal mound enclosed by a moat. An additional circular earthen embankment was recently constructed around the mound for conservation purposes. Recent road construction on the edge of the surrounding moat encountered a horizontal tunnel that begins at the enclosing moat and projects about five meters toward the central mound. The original extent of the tunnel is presumed to have continued into the central portion of the mound, leading to a burial chamber. Some test GPR profiles that were acquired across the surface features indicated that there may be a significant radar reflection generated at the discontinuity between the partially collapsed tunnel and the surrounding fill material (Fig. 16).

Information obtained from excavations at other round mounds similar to Kofun 111 indicated that there was a good chance for preservation of a stone chamber, with possible secondary burials within it. Nearby mounds that were excavated were found to contain shafts that were sometimes dug from outside the moats inward toward a central burial.

A GPR survey was conducted over the Kofun 111 mound in 1995 using dual 300 MHz antennas. A 50- by 50-meter grid was surveyed over the mound with lines collected every meter. Twelve time-slice maps were then constructed for the grid, with a two-way-time slice interval of 14 nanoseconds (Pl. 5a). Each slice represents approximately 56 centimeters in the ground, using an average velocity of about 8 centimeters per nanosecond. No corrections were made for surface topography.

The recently constructed moats surrounding the mound are easily visible in the top slice from 0 to 14 nanoseconds. The deeper slices from 14 to 42 nanoseconds show the remains of two original moats that surrounded the interior mound. From 56 to 70 nanoseconds (corresponding to between approximately 2.2 and 2.8 meters depth) the surface moats are no longer visible, and only features within the mound are imaged. Because no topographic corrections were made, the high-amplitude reflection anomalies outside the area of the mound were likely generated from undisturbed sediments not associated with the mound building. In the slice from 112 to 126 nanoseconds (corresponding to a depth of between approximately 4.5 and 5 meters), a relatively high-amplitude anomaly is visible at coordinates $x=3$, $y=30$ meters. This anomaly continues in a straight line to the north, ending in the middle of the mound. It was likely generated at the interface of an interior tunnel and the surrounding soil that leads to a central, deeper, stone chamber. This feature may connect to the tunnel that was encountered to the east (Fig. 16).

An interior chamber is visible in the three deepest slices, from 126 to 168 nanoseconds, corresponding to a depth from approximately 5 to 6.7 meters. The chamber anomaly was likely produced by velocity discontinuities between a central stone-lined chamber and the surrounding volcanic soil used in construction. Anomalies to the north of the mound in the 126 to 168 nanosecond slice may have been generated from features produced by quarrying operations during the construction of the mound, or they may only be natural structures in the surrounding native sediment that have no archaeological significance.

A very interesting linear feature is visible within the mound on the 56 to 70 nanosecond time slice, which corresponds to approximately 2.2 to 2.8 meters in the ground (Pl. 5a). This feature appears to have been generated by a buried tunnel or trench that is approximately 14 meters in length. Archaeologists familiar with other excavated mounds in the area believe that this anomaly may represent a second entrance into the central chamber, or possibly a secondary burial interred after the primary burial chamber was constructed.

The total time to complete this survey and produce the slice maps was four hours, with an additional three hours of computer processing time. No corrections were made for topography, which would have increased the field-work time substantially due to the detailed surveying necessary to obtain accurate elevations. If accurate surface surveys are available, there are computer programs available that can easily correct profiles and slice maps for elevation changes. In this case, they were not necessary because a knowledge of the internal characteristics of other nearby features allowed archaeological and natural anomalies to be differentiated.

THREE-DIMENSIONAL MAPPING

Three-dimensional mapping of anomalies with amplitude time slices can be accomplished quickly, especially when the ground surface is flat and no topographic corrections are necessary. This was performed in a GPR survey conducted in 1995 on a flat surface located on top of a large keyhole-shaped burial mound called Yamashiro Futagozuka, in Shimane Prefecture, Japan (Fig. 33). Test excavations discovered the presence of a stone-lined entrance to the main portion of the mound (Fig. 37) prior to the GPR survey (Archaeology Department, Shimane Prefecture 1978). The purpose of the survey was to determine the exact physical dimensions of a possible stone-lined chamber connected to this entrance within the main portion of the mound.

A 7- by 16-meter GPR survey was conducted using dual 300 MHz antennas, with a line spacing every 50 centimeters. Ten time slices were created, one for each 10 nanoseconds (Pl. 5b) from the surface to 100 nanoseconds (two-way time). Each slice represents approximately 60 centimeters in the ground. The slice from 40 to 50 nanoseconds (approximately 3.8 to 4.75 meters) indicates the presence of very high-amplitude reflection in the eastern portion of the grid (Pl. 5b). This anomaly is directly in line with the excavated entrance located on the eastern edge of the mound. The depth and location of this anomaly indicates it was likely generated from a stone-lined burial chamber. When connected to the known entrance to the east, the total length of the stone tunnel leading to the burial chamber is between 13 and 14 meters, making it one of the longest archaeological features of this sort ever discovered in this area of Japan. The length of the interior burial chamber, which is well-defined by GPR, could be precisely measured as 4.5 meters long and 2.5 meters in width. In progressively deeper slices, only reflections from the back wall of the chamber are imaged. In order to view the chamber in three dimensions, a cutaway section was constructed using the amplitude anomalies generated in all the slices (Pl. 6a).

The top of the burial chamber is estimated to be 3.8 meters below the surface because it was first visible in the 40 to 50 nanosecond slice. Any stone coffins that may exist within the chamber are not visible in the slice maps because most of the radar energy was reflected at the interface between the stone lining and the air within the chamber.

The GPR survey at the Yamashiro Futagozuka Mound was instrumental in defining the exact location and dimensions of the burial chamber. Excavations in the future can now be conducted only where the central chamber is located, and other portions of the mound can remain undisturbed.

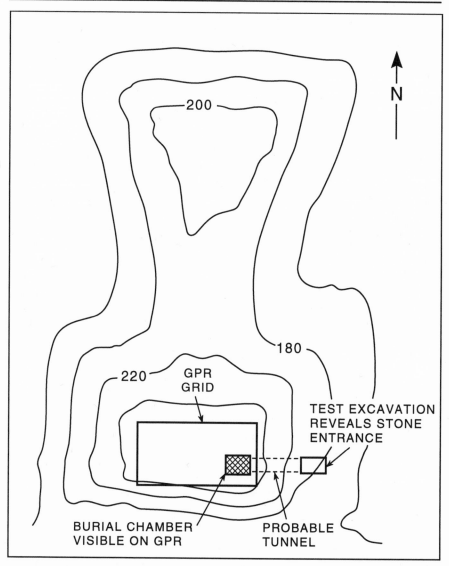

Figure 37. Surface topography at the Yamashiro Futagozuka Mound, Japan. The "keyhole" nature of the mound is clearly defined. Contours are in meters above sea level. A test excavation, which located a stone entrance to the interior chamber, is located east of the GPR grid. This entrance likely connects to the burial chamber discovered with GPR.

IDENTIFICATION OF FEATURES INVISIBLE IN TWO-DIMENSIONAL PROFILES

The Spiro Mounds, located in Oklahoma, represent the westernmost extent of significant mound building in North America during the Mississippian Period (Rogers et al. 1989). The earliest occupation of the Spiro Mound area was approximately A.D. 900; and the site was abandoned about A.D. 1450. The Spiro area was an important focus of trade in the region as well as a ceremonial and religious center. Trade to and from the Spiro area has been documented as far south as the Gulf of Mexico and north almost to the Canadian border.

A number of mounds are visible in the Spiro area, some of which have been excavated (Fig. 38). Mound 6 was chosen for a GPR survey because it was believed to have been left undisturbed by early-twentieth-century looters. Only a slight topographic rise remains after erosion, late-nineteenth-century plowing and land leveling, and some controversy existed about whether the mound was actually an archaeological feature. It was hoped that GPR might be capable of imaging internal anthropogenic features.

Excavations conducted prior to the GPR survey indicated the mounds have very little visible internal stratigraphy. Nearby excavations also indicated the remains of a number of burned buildings that were covered by earthen mounds. Some of the burned structures are thought to have been ignited after the death of a chief or religious figure, possibly during funeral ceremonies.

Mound 6, which is visible as a small topographic feature (Fig. 38), was surveyed by GPR in 1992. About a meter of topographic relief presently exists over the area where the GPR survey was conducted. Dual 300 MHz antennas were used to acquire data within a 33 by 39 meter grid, with a one-meter line spacing. Velocity analyses were conducted that arrived at an average radar-wave velocity of about 6 centimeters per nanosecond.

Topographically uncorrected two-dimensional profiles showed little in the way of buried features of interest (Fig. 39). The topography of the grid was then surveyed in order to make surface elevation corrections so that amplitude slices could be made parallel to any possible horizontal stratigraphy that might exist within the mound. All reflection traces in each GPR profile were then corrected using survey values obtained every 50 centimeters along grid lines, assuming a velocity of 5 cm/ns across the site (Fig. 40).

Soils at the site are very fine-grained and contain abundant clay that may have attenuated much of the radar energy below about 40 nanoseconds. Initial viewing of the raw profile data in the field was discouraging because there were few reflections visible that might have delineated important archaeological features.

Horizontal time slices were produced for the grid after each line was corrected for surface topography. In this way, reflections from possible internal stratigraphy within the mound could first be reconstructed for surface

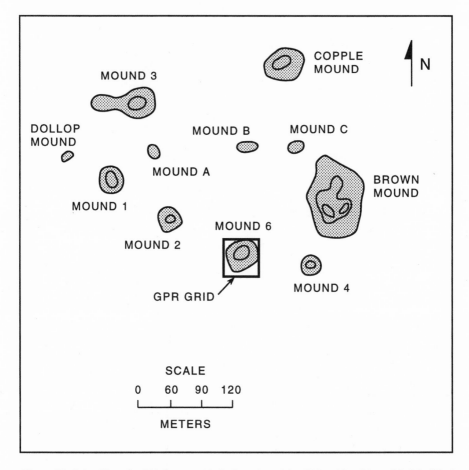

Figure 38. Spiro Mounds, Oklahoma, with the location of the GPR grid on Mound 6. (Modified from Rogers et al. 1989.)

Om   **34m**

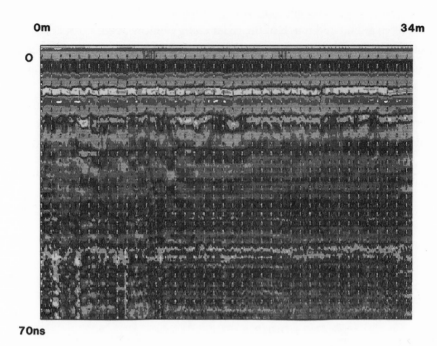

70ns

Figure 39. Representative two-dimensional profile across Mound 6, Spiro Mounds, Oklahoma. This profile is uncorrected for surface topography, and the background is not removed. Little in the way of buried archaeological features is visible in this profile.

topography and then sliced horizontally, parallel to any possible internal bedding planes.

Horizontal amplitude slice maps were constructed every 7 nanoseconds (Pl. 6b). The first slice, from 0 to 7 nanoseconds, illustrates no major amplitude anomalies within the slice located from the ground surface to about 40 centimeters. Because all traces were corrected for topography, this uppermost map includes reflections only within the top of the topographic rise. Each consecutively lower horizontal slice cut deeper into the mound. Little can be seen in these upper slices, down to about 35 nanoseconds, due to the lack of internal stratigraphy or any burial features. In the deepest slice, from 35 to 42 nanoseconds (about 85 to 100 centimeters in depth) a pronounced anomaly about 20 meters square is visible. This anomaly likely represents the foundation and walls of one or more Mississippian buildings that may

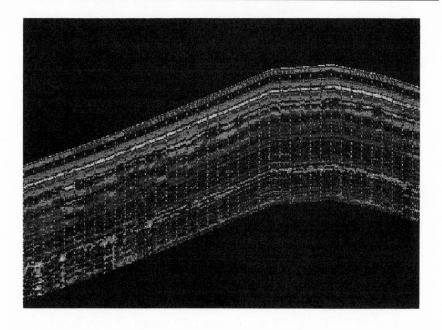

Figure 40. The same two-dimensional profile in figure 39 after it was corrected for surface topographic changes.

have been periodically burned and then rebuilt. The building was covered over in prehistoric times during construction of the mound.

Amplitude slices that were constructed prior to correcting for topography showed no rectangular anomalies in any of the slices due to slices crossing bedding planes. At this site, the archaeological features would not have been visible if both topographic corrections and amplitude analyses had not been performed. The survey would likely have been deemed a failure due to the lack of visible reflections in the two-dimensional profiles. And in this case, even if time slices had been made, but no topographic corrections done, success would still have eluded the surveyors.

It is sometimes possible to correct for topography without actually having to make detailed surface surveys. If a subsurface reflection is visible on profiles, and it is known that the layer being imaged is relatively flat or has a specific measurable slope, then all reflection traces in two-dimensional profiles can be normalized to its elevation. In this process, each trace within profiles is arbitrarily flattened on the computer during processing. To accomplish this, the reflection to which all traces are being normalized must first be identified on all profiles, and the time it was recorded noted. This is best accomplished by viewing each line on a computer screen and recording the reflection's location within all traces with the aid of a hand-operated computer mouse. That reflection is then arbitrarily assigned a two-way time, and the digitally recorded data from all other traces in the profile are adjusted up or down relative to that value.

HORIZON-SLICE MAPS

All of the amplitude analysis cases shown so far in this chapter have focused on the production of horizontal slices. These maps are satisfactory when the buried features or stratigraphy are also horizontal, or nearly so. If the buried features are horizontal but appear distorted due to uneven surface topography, horizontal slices can still be produced if topographic corrections are made before a grid is sliced. However, horizontal slices may not be suitable if there is a great deal of subsurface complexity. For instance, if the stratigraphy or archaeological features within a grid rise and dip dramatically, horizontal slices will cross bed boundaries often, creating false amplitude anomalies wherever the slicing and bedding planes intersect. In order to overcome this problem, slices must be prepared that can parallel bed boundaries of interest, rising and falling with the layers instead of arbitrarily crossing them. These types of slices, first used in seismic exploration, are called horizon slices.

Horizon slices are constructed by viewing each profile in a grid on the computer screen and using a mouse to "draw" a time window along a reflection or series of reflections of interest. This is done with the aid of an interactive computer program. Care must be taken when choosing the slicing window thickness so that the horizon of interest is always included within the window. The amplitudes of the reflected waves within a varying-depth window are then input into a separate file, where they can be analyzed for contrasts.

Horizon slices were produced from data acquired at the Asada site in Komochi Mura, Guma Ken, Japan (Fig. 33). At this site, possible pit dwellings and a living surface are preserved, buried beneath more than one meter of volcanic ash (Isseki 1993). A schematic drawing of a nearby pit dwelling that has been excavated is shown in figure 41. When the pit dwellings in this area were constructed, they were first dug about a meter into the volcanic soil. Support timbers were then placed so that the roofs could be covered with soil and sod. A nearby volcanic eruption buried the site in about A.D. 1000, collapsing the roofs and filling the pits of many of these structures. The volcanic material that covered the site was slightly different in chemistry, porosity, and density than the existing materials that made up the pit floors and the surrounding ground surface, creating a velocity contrast at the interface, which reflects radar energy.

A 20- by 35-meter GPR grid was acquired at the site using dual 300 MHz antennas. Good reflections were visible to about 50 nanoseconds, which is approximately 3 meters deep (Fig. 42). Initial examinations of the two-dimensional reflection profiles indicated the presence of a possible depression, within which a pit dwelling might exist. One depression, visible on a number of profiles, was approximately 90 centimeters deep and about 20 meters wide. Standard horizontal time slices that were first conducted every 5 nanoseconds across the grid proved to be unsuccessful because the slices crossed many stratigraphic layers.

Within the GPR grid, all east-west lines were viewed on the computer screen and a time window of 15 nanoseconds was overlaid on the high-amplitude reflection that was created at the contrast of the ancient living surface and the overlying volcanic ash (Fig. 42). The top of the horizon slices that were chosen for each of the profiles in the grid are shown in figure 43.

An amplitude anomaly map of the horizon slice was then constructed in the same way as standard horizontal slices (Fig. 44). The resulting anomaly map shows a very distinct low-amplitude anomaly at 5 to10 meters east and 12 to 20 meters north. This feature is somewhat rectangular in shape and is about the same dimension as some of the nearby pit dwellings that have been excavated. When viewed in three-dimensions, the location of this low-amplitude feature on the ancient landscape becomes apparent (Fig. 45). The low-amplitude anomaly that is visible on the maps indicates a lack of significant reflection in the area of the possible pit dwelling. This lack of reflection is probably due to a lack of contrast within the volcanic material that filled the interior of the pit. The actual pit floor itself, and any reflections that may have been derived from the interface of its floor and the overburden, is not

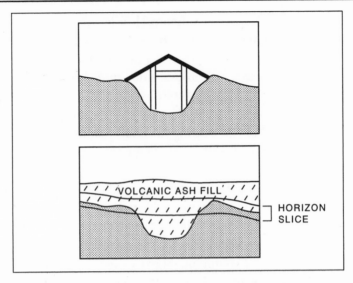

Figure 41. Diagrammatic cross section across a pit dwelling, Asada Site, Japan. In the upper drawing, the pit dwelling is shown as it was originally constructed. In the lower drawing, the roof has collapsed and the pit filled with volcanic ash. On the computer, a horizon slice was constructed along GPR profiles that followed the reflection generated at the ancient living surface and crossed the pit fill, but not the floor of the dwelling.

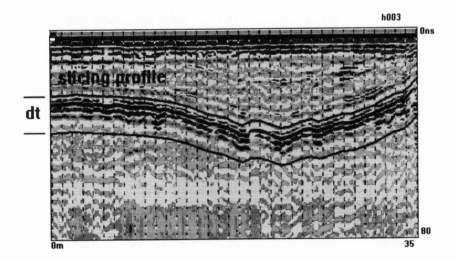

Figure 42. Two-dimensional profile across the Asada Site pit dwelling, with the location of the horizon slice, shown as d_r.

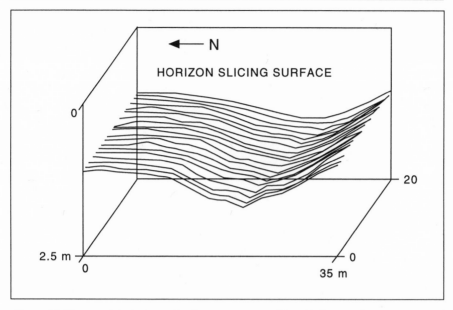

Figure 43. The top of the horizon-slice layers constructed for all GPR profiles within the grid at the Asada Site, Japan.

part of this horizon slice (Fig. 41). The floor is not readily visible in any of the standard reflection profiles below the horizon slice, possibly due to attenuation of the radar energy below about 3 meters. The other amplitude changes away from the pit-dwelling anomaly that are visible in the horizon slice in figure 45 probably represent changes in soil types or anthropogenic modifications of the ancient living surface prior to its burial by ash.

Horizon-slice maps are very beneficial when studying changes on a specific surface that is located at varying depths. If a surface that is preserved as an interface between two contrasting materials can be identified as one radar reflection on profiles, amplitude changes that are distinct to only that horizon can then be isolated and studied. Because amplitude anomalies are caused by changes in the reflectivity at an interface, it is important to know the physical characteristics of the material both above and below the interface that may be causing the reflection. If the material overlying the boundary of interest is fairly homogeneous, then most of the amplitude changes visible on the anomaly maps

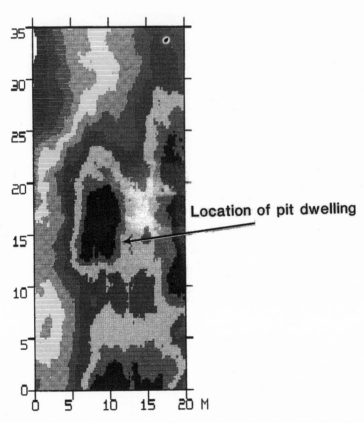

Figure 44. Amplitude anomalies produced from the three-dimensional horizon slice surface, Asada Site, Japan. The dark anomaly, indicating little reflection in the pit's fill material, represents the location of the dwelling. (After Goodman et al. 1995.)

will be related to changes in the material at, or directly below, the interface. If, however, there were large changes in the materials both above and below the interface, many of the amplitude anomalies would be very difficult to interpret unless a great deal was known about both the subsurface stratigraphy and the physical characteristics of both materials.

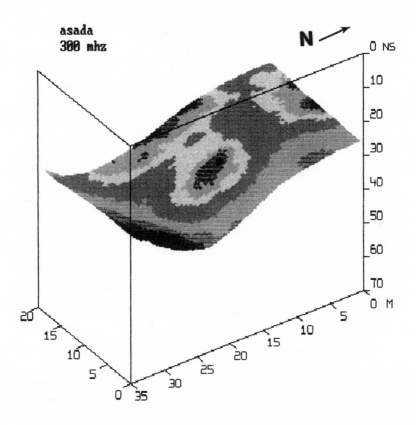

Figure 45. Three-dimensional amplitude anomaly map produced from the horizon slices shown in figure 43, Asada Site, Japan.

Horizon-slice interpretation can be complicated somewhat if the horizon that is being sliced changes its depth dramatically across a grid. When depths change, amplitude changes of one reflection may be related to differences in attenuation at the variable depths along profiles rather than meaningful physical changes at the interface. Another complicating factor in the interpretation of horizon-slice maps is related to the resolution of buried interfaces in the subsurface. Long-wavelength radar-energy reflections are usually not capable of resolving the details of thin beds. If low-frequency antennas are used to collect the data being used to produce the slice maps, then the amplitude differences derived from one distinct reflection may be illustrating gross changes over a thick section of material. If this is the case, amplitude changes may have more to do with large-scale lithologic or water-saturation changes over a thick section of material than with features of archaeological interest. Radar-energy focusing and dispersion caused by variations in material velocity (Fig. 5) or the scattering effects caused by the orientation of subsurface interfaces (Fig. 12) can also create amplitude variations. These changes are many times difficult to interpret, and amplitude anomalies caused by them can be confused for archaeologically or geologically significant changes.

In addition, if the resolution of subsurface units is good but the ground water table or residual water saturation varies at the boundary of interest, large differences in reflectivity, and therefore amplitude, will be created. Unless these changes can be directly related to changes in physical properties along the bed boundary, they can be confused with anthropogenic or natural changes along the buried surface when horizon-slice maps are created.

INTEGRATION OF GPR AMPLITUDE DATA WITH RESISTIVITY AND MAGNETIC MAPS

The Matsuzaki site in Chiba Prefecture, Japan (Fig. 33), provides a good example of how GPR data can be used to identify areas that have previously been excavated, as well as discover other nearby archaeological features that may have been missed. In addition to GPR, resistivity and magnetic gradiometer surveys were performed at this site that allow a comparison of the GPR results to other geophysical methods.

The Matsuzaki site was excavated in 1983 by the Educational Department in Chiba (Hiroshi 1989). Two parallel exploratory trenches were excavated across the site and were later backfilled. During these initial archaeological operations, portions of two pit dwellings from the Final Jomon Period (about 2500 B.C.) were discovered within 40 centimeters of the ground surface. Only the edges of the pits were uncovered during excavations, and the floors were left undisturbed (Figure 46). A Medieval ceramic kiln was also discovered nearby during a later excavation in 1992.

A 30- by 40-meter GPR grid, with a one-meter line spacing, was acquired across all the known archaeological features in 1993. Reflection data from 30 north-south lines were obtained using dual 300 MHz antennas. An example of one of the profiles that was located across one of the known pit dwelling is shown in figure 47. This 14-meter segment of a longer profile depicts a high amplitude reflection between 30 and 40 nanoseconds that was generated at the interface of the pit dwelling floor with the overlying soil.

Horizontal-amplitude time slices were constructed from the ground surface to 42 nanoseconds (Pl. 7a). Each time slice represents a depth of approximately 21 centimeters in the ground. The uppermost slice, from 0 to 7 nanoseconds, displays anomalies representing two different surface soil characters that are the result of different agricultural practices on either side of a north-south property line. The low-amplitude, dark blue anomaly in the southeastern portion of this upper slice denotes the location of a recently dug garbage pit.

The third slice, from 14 to 21 nanoseconds, depicts anomalies in a layer of soil that is above the horizons of archaeological interest. The anomalies generated in this and the overlying slice may be the result of soil mixing during plowing operations within the last ten years.

In the slice from 21 to 28 nanoseconds, two parallel linear features are clearly visible, trending west-northwest. These anomalies represent the test trenches from the 1983 excavation that were backfilled with less compacted soil than the surrounding units. A comparison of these anomalies with the map derived from the 1983 excavation (Fig. 48) shows a good correlation between the trench locations and these anomalies. They are not visible in the overlying slice due to the homogenization of materials that occurred during the plowing that followed excavation.

In the slice from 28 to 35 nanoseconds, six distinctive round features are visible (Pl. 7a). Two of these features are known pit dwellings discovered during the 1983 excavations. The linear features created by the backfilled trenches are still visible in this slice. A comparison of the location of the small circular features to the excavation map (Fig. 48) demonstrates that the high-amplitude feature in the northwestern corner of the grid was generated from a Medieval ceramic kiln. After shallow field tests, the other three round anomalies visible in the 28 to 35 nanosecond slice were discovered to be a modern garbage pit, the ceramic kiln, and a previously unknown pit dwell-

Figure 46. Excavation of a pit dwelling at the Matsuzaki Site, Japan. The soil discoloration, denoted by the lines drawn on the excavation surface, approximates the outline of the wall of the pit dwelling. (Photo courtesy of Y. Nishimura.)

ing. These three features and the two known pit dwellings are also visible on the deepest slice, from 35 to 42 nanoseconds, which represents a horizontal slice in the ground from about 105 to 126 centimeters. The test trenches, which were not excavated to this level, are no longer visible in this deepest slice.

The GPR data displayed in time-slice maps and the magnetic gradiometer and two-probe resistivity data obtained as a part of the same geophysical program make for an interesting comparison of geophysical methods. The magnetic gradiometer map (Fig. 49) displays measurements of differential changes in the earth's magnetic field that are caused by changes in the properties of near-surface soils or other features (Scollar et al. 1990). Magnetic gradiometers use two sensing devices mounted on a vertical staff. The sensitivity between the devices can be calibrated so that its maximum sensitivity is within the near-field zone, which is the area closest to the ground surface.

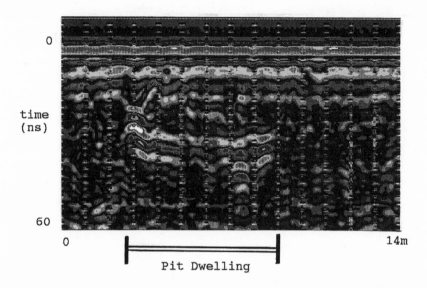

Figure 47. Two-dimensional GPR profile across a pit dwelling, Matsuzaki Site, Japan. The floor of the pit dwelling, at approximately 30 nanoseconds, is visible by a dark-colored, high-amplitude reflection.

Magnetic variation readings are then obtained across the grid and the anomalous zones are plotted on a map similar to amplitude anomalies in GPR.

The plotted magnetic anomalies, obtained across the same grid from which the GPR was acquired, were capable of defining the southern test trench, one of the known pit dwellings, the modern garbage pit and the Medieval kiln (Fig. 49). The second test trench and the pit dwelling in the eastern portion of the grid, visible on GPR maps, are not visible on the magnetic map. The magnetic data did, however, discover many of the same features that the GPR amplitude slices did, with a minor reduction in clarity. Some of the features that are well-defined by GPR, however, were not visible in the magnetic anomaly map, probably because they had no magnetic contrast.

A second geophysical test using a resistivity meter in a twin-probe configuration was also conducted over the same grid (Fig. 49). This test uses two conductive probes that are placed into the soil at a set separation distance. An electrical current is then delivered to one probe, and the amount of current received at the other probe is measured. The conductivity (and its inverse resistivity) of the soil and material within the field of the current can be directly measured and the values plotted on a map (Scollar et al. 1990). The greater the horizontal separation of the electrode probes, the greater the depth of electrical energy penetration.

At the Matsuzaki site, a 12-volt battery was used to deliver the electrical current into the soil between electrodes spaced one meter apart. The resistivity measurements obtained in the field were then plotted in map form (Fig. 49). None of the features that were visible in the GPR or the magnetic maps is visible in the resulting electrical resistivity map. The failure of this method to

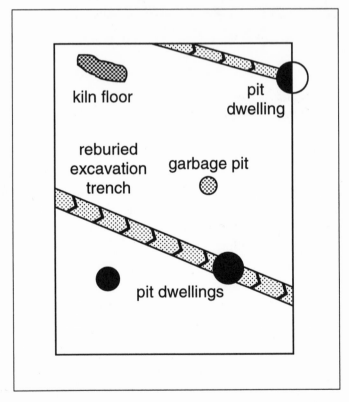

Figure 48. Excavation map of the Matsuzaki Site, Japan. All features on this map are also visible on the amplitude slice maps in plate 6a.

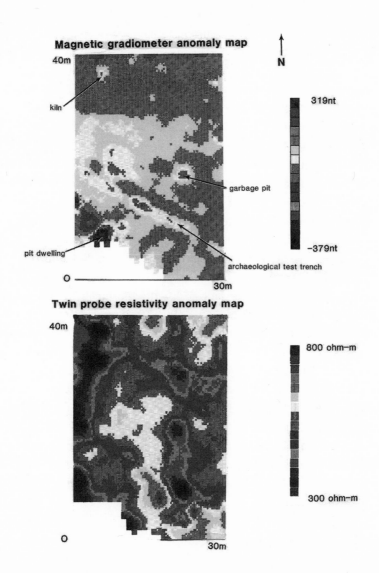

Figure 49. Magnetic gradiometer and twin probe resistivity maps, Matsuzaki Site, Japan. The grid used in these two geophysical surveys approximates those in plate 7a and figure 48. The magnetic map images some of the same subsurface features as the amplitude slice map produced from GPR reflection data. (Maps courtesy of Yasushi Nishimura, Nara National Cultural Properties Research Institute.)

discern any of the important features may be the result of electrodes that were spaced too far apart, allowing the current to pass too deeply into the ground. The resistivity measurements plotted on the map were therefore likely obtained from units that were deeper than the zones of archaeological interest.

Buried ceramic kilns such as that discovered at the Matsuzaki site are readily visible with GPR due to the reflective nature of the interfaces between fired-clay linings and the surrounding soils. A series of buried kilns similar to that described above was discovered in Suzu City on the Noto Peninsula of Japan (Fig. 33) by Goodman and others (1994) using GPR. In this area, a series of chambers including a central fire box and air intake and exhaust vents were usually dug into hillsides (Uno et al. 1993). Archaeologists suspected that kilns were located in this area because of the high number of ceramic sherds on the ground surface. The area had been used for growing crops for many years, and all other surface indications of possible buried kilns had been destroyed.

A 20- by 13-meter GPR data grid, with one-meter line spacing, was acquired using dual 300 MHz antennas (Fig. 50). Standard two-dimensional profiles of the resulting data showed reflection anomalies that likely were derived from three intact kiln floors (Fig. 51). When amplitude-slice maps were constructed, the kilns were readily visible in the slice from 12 to 16 nanoseconds (Fig. 50). The eastern two kilns were also visible as magnetic anomalies (Sakai et al. 1993). Excavations in the eastern portion of the GPR grid confirmed the location of the two eastern kilns (Fig. 52), but the presence of the western kiln, imaged by GPR but not by the magnetometer, is still speculative. It may be a kiln that was never used and therefore contains no baked edges that would yield a magnetic anomaly.

THE USE OF AMPLITUDE TIME SLICES TO SEARCH FOR VERTICAL FEATURES

Ground-penetrating radar is usually a good tool to explore for and map horizontal or planar features, because these types of broad buried discontinuities tend to reflect the most amount of radar energy. When the target features are both vertical and small in dimension, they can be very difficult to identify in standard two-dimensional profiles and are often overlooked. In an attempt to image vertical features, a series of amplitude time slices that were uncorrected for topography were constructed over two burial mounds where vertical shafts were suspected.

The Kofun 102 and 103 Mounds are located in the Saitobaru Park near the Kofun Mound 111 (Fig. 33). Some other small mounds in the area, similar to those at 102 and 103, are known to have unusual vertical shafts leading to burial chambers (Hongo Hiromichi, personal communication 1993). The two mounds were surveyed in 1993 using dual 300 MHz antennas (Fig. 53). A 20- by 45-meter grid was acquired with transects spaced one meter apart. In plate 8, eight time slices are shown at 8 nanosecond intervals. None of these slices is corrected for surface topography and as a result the uppermost slice is illustrating the soil characteristics within both the mounds and the surrounding soil units, as shown in figure 54. When reflections are not corrected for topographic variations, a "horizontal slice" is not really horizontal but will parallel the ground surface.

The uppermost slice, from 0 to 8 nanoseconds, corresponds to a depth from the ground surface to approximately 25 centimeters. Concentric circular features are visible as amplitude anomalies that represent a different construction material in the mounds than that located at a similar depth in the surrounding soil (Pl. 8). In the slice from 32 to 40 nanoseconds, four high-amplitude anomalies are visible, two in each mound. This slice is analyzing reflections from a depth between one meter and 1.25 meters. The locations of these four anomalies are oriented in the same direction within each mound. Anomaly A in Mound 102 corresponds to the exact location of a vertical shaft that was discovered during test excavations prior to conducting the GPR survey. The placement of the shafts on the mounds indicates that they were probably dug after the mounds were constructed and likely lead to subterranean burial coffins. Anomaly A was probably produced by the velocity contrast that occurred at the interface of the shaft with the surrounding mound fill material.

The three additional anomalies also visible in the slice from 32 to 40 nanoseconds are likely shaft entrances similar to A. The central burial chambers of Mound 102 are visible in the lowest slices, from 48 to 64 nanoseconds, as distinct circular anomalies. A representation of the orientation of the surface mounds, the shafts, and the possible burial tombs is shown in figure 55.

At this site, the time-slice maps were capable of imaging vertical features that would not usually be readily visible on standard two-dimensional

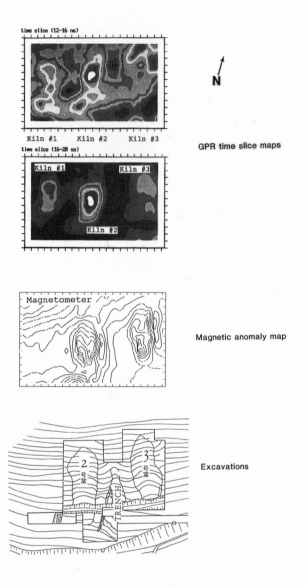

Figure 50. Amplitude slice maps of buried kilns, Suzu City, Japan. The two amplitude anomalies in the eastern portion of the grid (Kilns 2 and 3) have been confirmed by excavation and are also visible in the magnetic anomaly map. The western anomaly (Kiln 1) has not been confirmed by excavation and is not visible in the magnetic map. (After Goodman et al. 1994. Magnetic survey courtesy of Hideo Sakai, Dept. of Earth Science, Toyama University. Map after Sakai et al. 1993.)

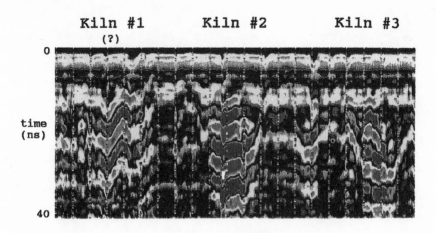

Figure 51. Standard two-dimensional GPR profiles across the confirmed and possible kilns imaged in figure 50, Suzu City, Japan.

Figure 52. Photograph of the excavations of the two eastern kilns, Suzu City, Japan. The baked clay floors of the kilns are visible as the darker-colored units. (Photo courtesy of Takao Uno, Toyama University.)

Figure 53. Collecting 300 MHz GPR data over the top of Kofun Mound 102, Saitobaru Park, Japan.

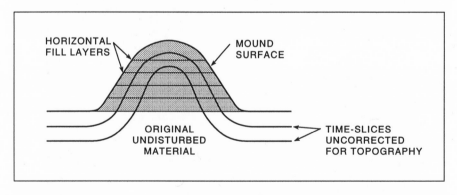

Figure 54. Diagrammatic cross section of uncorrected time slices. In an area with horizontal stratigraphy, two-dimensional radar profiles that are not corrected for topography will intersect layers of interest. When this occurs, any one time slice must be interpreted with caution because it is comparing amplitude anomalies derived at the intersection of the slices with bedding planes.

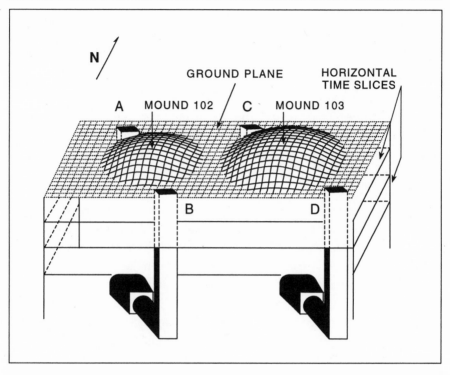

Figure 55. Representative diagram of the Kofun 102 and 103 Mounds and their vertical shafts leading to burial chambers. The vertical shafts and burial chambers are visible in the horizontal slice maps in plate 8.

GPR profiles. Two standard profiles across one of the features indicate the presence of an anomaly that might represent a vertical shaft (Fig. 56). Vertical features are usually difficult to notice in standard two-dimensional radar profiles, and may only be visible after reflection amplitudes are analyzed by the computer.

THE USE OF AMPLITUDE TIME SLICES TO IMAGE FEATURES IN THE NEAR-FIELD ZONE

Most researchers who use GPR believe that little, if any, useful reflection data are recorded within the near field zone. As discussed in chapter 2, this zone of poor or absent reflection data is located within about one wavelength of the ground surface. It is the zone within which transmitted radar energy is in the process of coupling with the ground, and therefore little reflection is occurring. It is usually displayed in two-dimensional GPR profiles as a hazy near-surface band, often incorrectly referred to as the "zone of interference" (Fig. 13). Due to the wide bandwidth of most GPR antennas, however, there can be some reflection of radar energy within this zone from the shorter wavelengths in the higher-frequency portions of the transmitted band. Shorter-wavelength radar energy will couple with the ground nearer the surface and is therefore available for reflection, while the longer-wavelength energy is still in the process of coupling deeper in the ground. Unfortunately, in most standard two-dimensional profiles, the dominance of the longer wavelength band tends to obscure any near-surface reflections of the shorter wavelengths.

Amplitude analysis on the computer has the ability to extract the short-wavelength reflected signal from the background noise and display it in a form that is useful to the interpreter. Amplitude anomaly maps can then display the extracted amplitude anomalies that may be represented by only a few digital values in the recorded waveform. Using this technology, recorded data that were formerly ignored can now be manipulated in order to answer important archaeological questions.

A good example of this near-surface extraction process is from GPR data recorded at the Shawnee Creek site in Eminence, Missouri. Several Mississippian Period (about A.D. 1000) fire pits and one structure, which had been burned prior to abandonment, were discovered during excavations by James Price of the University of Missouri and Mark Lynott of the National Park Service. All open pits and other excavations were later backfilled. In an attempt to discover other possible structures and archaeological features in the same area as the excavations, a 20- by 60-meter GPR survey was conducted over the area in 1992. All lines within the grid were spaced one meter apart.

Initial analysis of the two-dimensional profiles that were recorded in that survey indicated that the high clay content of the soils and sediment in this area might have limited radar-energy penetration to only about 50 nanoseconds (Fig. 57). Little of interest appeared as reflections or visible amplitude changes in any of the profiles. Even those data that were collected directly over the known location of the shallow burned structure showed no anomalous reflections corresponding to either the archaeological feature or the backfill area, due to the

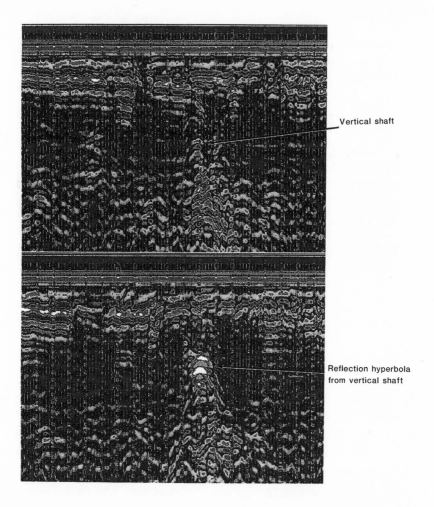

Figure 56. Parallel two-dimensional GPR profiles across a vertical shaft, Kofun Mound 103, Saitobaru Park, Japan. The vertical shaft is barely visible in the upper profile, but it is more obvious as a reflection hyperbola in the lower profile.

near-field effect. Prior to the use of amplitude time-slice analysis, profiles such as those in figure 57 would probably have been deemed worthless and the survey would have been considered a failure.

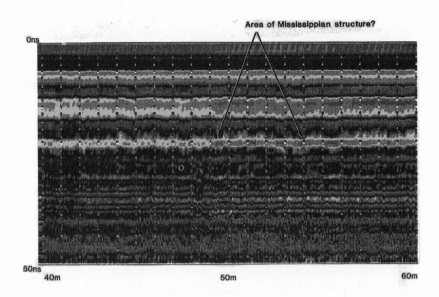

Figure 57. Two-dimensional GPR profile, Shawnee Creek Site, Missouri. Background noise has not been removed. There are no visually interpretable reflection data either within the near-surface zone or deeper within this profile. Data of these sort, when processed using amplitude slice maps, can produce meaningful three-dimensional anomalies.

In an attempt to extract some useful information from these data, amplitude time slices were constructed every 5 nanoseconds. Two-way travel times were then converted to depth, using a velocity of 6 cm/ns, and slices of amplitude anomalies were printed every 15 centimeters. A very-high-amplitude anomaly was discovered in the northwest portion of the grid in all slices from the ground surface to a depth of about 75 centimeters. This anomalous area, designated Feature 1, represents the excavated area where the burned Mississippian structure was discovered (Fig. 58). The anomaly was probably generated by reflection amplitude changes generated in the relatively uncompacted soil used to backfill the excavations. The location of other nearby excavations that were also conducted during the 1980s are overlain on the shallow time-slice map (Fig. 58). There is a good correlation between these shallow anomalies and excavations. A series of horizontal amplitude slices was also constructed at approximately 15 centimeter intervals to a depth of 150 centimeters (Pl. 7b).

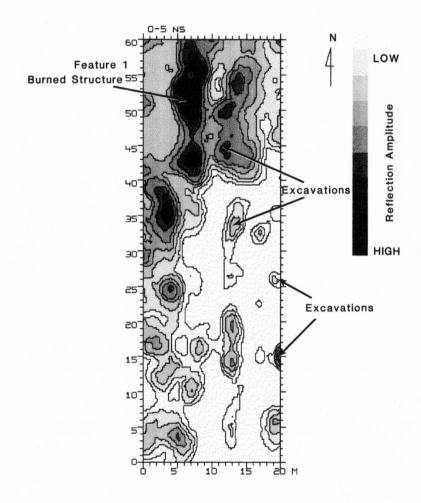

Figure 58. Uppermost time-slice from the Shawnee Creek Site, Missouri, with the location of the excavations. There is a good correlation between the location of the archaeological excavations and the amplitude anomalies, demonstrating that there are usable data still recorded within the near-surface zone that are not visible in standard two-dimensional profiles like that shown in figure 57.

In the deeper slices, from 75 to 150 centimeters, a number of amplitude anomalies that were not visible in standard two-dimensional profiles are visible that continue over many slices. They likely represent additional buried structures similar to the burned house that was excavated. Only through the use of a computer to extract the important anomalous reflections are these features visible.

Chapter 9

Conclusions

Many of the case studies used to illustrate GPR techniques and interpretative results in this book were "best cases," obtained by the authors from many surveys conducted under differing conditions in diverse environments. We have also conducted other surveys that were either not as successful or totally failed; these of course are not presented. Our failures and marginal successes highlight the fact that GPR exploration is not a panacea for all archaeological problems confronted in the field. We therefore assume some risk in presenting only the best cases, leaving readers with the impression that GPR exploration can automatically be applied to their own problems. This may not always be the case.

It is important to emphasize that archaeologists cannot arbitrarily employ the GPR techniques presented in this book without having to break the ground with a shovel and get on their hands and knees with a trowel and dustpan. GPR will never be able to replace standard archaeological methods, and for it to be most successful, the method should be integrated with them. Subsurface radar reflections will never be able to determine things like the age of an archaeological feature, what kind of pottery it may have in context with it, or the color of the pigment it is decorated with. Only excavations can yield this type of information. The method's strength lies in its ability to discover hidden features, create accurate images of them in three-dimensions, and produce maps of important stratigraphy between and surrounding standard archaeological excavations.

In its current form, ground-penetrating radar is capable of discovering otherwise invisible features quickly and accurately over broad areas. This can be done inexpensively and non-destructively compared to standard excavation methods. A typical 50-meter-square grid, with surface antenna transects spaced one meter apart, can be acquired in about four hours. The data can usually be processed the same day if a portable computer is taken to the field. After a survey is finished, the land surface has been marred only by the superficial marks of radar sleds being pulled over the ground and by footprints.

Many in the archaeological community will continue to employ GPR in its traditional role as an "anomaly-finding" device to locate features that can later be excavated. Although this will remain an important usage in the future, we believe that GPR achieves maximum effectiveness when it can be integrated with detailed archaeological and geological information collected from excavations and stratigraphic studies. When this is done, the method becomes a valuable tool to accurately map the anthropogenic and natural environment of a site. Without knowledgeable input from experienced archaeologists and stratigraphers, GPR is nothing more than a "black box" method that creates interesting (and possibly important) two-dimensional images of features under the ground. Its strength lies in its integration with all other methods of archaeological field research.

It must be remembered that the GPR method is only effective as a mapping tool in certain environments, under specific soil and moisture regimes. Its success or failure is premised on the knowledgeable application of the correct equipment, the appropriate input parameters, and the interpretative ability of the archaeologist, geologist, or geophysicist. It is one matter to recognize GPR anomalies (which may or may not have significance) and quite another to interpret the most important reflections correctly in order to derive archaeological or environmental meaning.

In an attempt to specify some of the successes and failures of GPR, an assessment of the feasibility of imaging various features in the ground is presented in table 6. These qualitative assessments are derived from the experiences of the authors alone, and it is hoped that others using slightly different techniques may produce better results for some of what we consider the more marginal archaeological applications. In addition, we admit that our own GPR experience has been focused on only a very small fraction of the sites and conditions around the world where the method could potentially be applied. Each of the studies used as illustrations in this book was made during specific times, under situations that can never be replicated due to constantly changing environmental conditions. Table 6 must therefore be used only as a general guide for accessing the feasibility of GPR; it cannot be relied upon as an exclusive guide to all situations.

Table 6

Feasibility of Using GPR to Discover and Map Buried Archaeological Features and Stratigraphy

Archaeological Target	Feasibility for GPR	Reasons for Assessment
Pit dwelling filled with different material than the surrounding matrix	Good	Good velocity contrast between floor and matrix will produce strong reflections
Trenches, buried moats, hollow tunnels	Excellent	Good velocity contrasts at interfaces of features with surrounding material of the void
Reburied excavation trenches	Moderate to good	Good velocity contrast when backfill material is different than surrounding materials or is less compacted
Fire pits with baked bottoms greater than 1–2 meters in diameter	Good	Firing can create a baked surface which reflects radar energy
Fire pits less than one meter in diameter	Moderate to poor	Usually too small to be visible unless pits are very shallow and high-frequency antennas are used
Stone foundations buried in fine-grained material	Good	Vertical walls will create reflection hyperbolas from their tops and sides

(continued on next page)

Table 6—Continued

Archaeological Target	Feasibility for GPR	Reasons for Assessment
Clay, stone, or wooden structures buried by rocky material	Poor	Too many small reflections (clutter) are created from the rocky material
Clay, stone, or wooden structures buried by wet clay	Poor	The high electrical conductivity of clay will severely attenuate radar energy
Clay, stone, or wooden structures buried by moist or dry fine-grained volcanic material	Good	Good velocity contrast between the structures and the surrounding material creates strong reflections
Kiln floors or roofs	Excellent	The high temperatures in the kilns baked the surrounding material, creating an excellent radar reflection surface
Buried living surfaces overlain by material of a different lithology	Good	A spatially extensive interface with a good velocity contrast will reflect radar waves
Small stone tools dispersed in soils	Poor	Target objects too small to reflect enough radar energy to be visible
Small metallic tools dispersed in soils	Moderate	If objects are not buried too deeply and high-frequency antennas are used, metal objects will create small, but visible reflection hyperbolas; can also create very visible multiple reflections

Table 6—Continued

Archaeological Target	Feasibility for GPR	Reasons for Assessment
Moderate to large metal objects	Excellent	Metal is a perfect radar reflector and will generate visible reflection hyperbolas
Small clay artifacts	Poor	Usually do not create a large enough velocity contrast with surrounding material to be visible as a reflection
Spatially extensive concentrated areas of pottery sherds	Moderate	Depending on their thickness and velocity contrast with the surrounding matrix, they can create a noticeable reflection that appears as one layer
Burials filled with material that is different than the surrounding matrix	Moderate	Good velocity contrast at the interface of the two different materials may create good reflections; visible if burial is large and not too deeply buried
Rock or clay-lined burials in fine-grained matrix	Good	Rock and clay usually make a good contrast with the surrounding material, and burials are usually large enough to be visible with most radar frequencies
Features within rock-lined chambers	Poor	Cap stones and rock lining will reflect most of the radar energy before it can enter the chamber

(continued on next page)

Table 6—Continued

Archaeological Target	Feasibility for GPR	Reasons for Assessment
Compacted mud or soil walls and floors buried by fine-grained material	Good	Good velocity contrast exists at the wall interfaces, which will create reflections
Stratigraphic layers with thicknesses less than the transmitted radar wavelength	Poor	The top and bottom of the layer cannot be resolved because the wavelength of the transmitted energy is too long
Stratigraphic layers thicker than the transmitted radar wavelength	Good	Both the top and bottom of the layer will reflect energy from the same transmitted waves
Any feature below a thick, wet clay layer	Poor	The thick wet clay is very electrically conductive and will attenuate most, if not all, radar energy

The feasibilities described in table 6 must be tempered by the understanding that the depth of the targets, as well as local ground conditions, can drastically alter radar-energy penetration and reflection. A common misconception is that resolution of buried archaeological features or important stratigraphy is better near the surface and decreases with depth. Generally this is true, but in some cases the antenna's near-field zone will obscure important shallow reflected waves that can only be coaxed out of the data base with sophisticated computer analyses. In addition, if the profile spacing in a grid is too large, some shallow features lying between antenna transects will remain undetected no matter how good the subsurface resolution may be. These same features, however, if buried deeper in the ground, might be visible with the same line spacing due to the conical spreading of the radar beam that "illuminates" an ever-increasing area with depth.

Many believe that GPR surveys can only be conducted on level or nearly level ground. It is always easier to work on a clear flat surface, but uneven or rough ground should not preclude using GPR. Some successful surveys, conducted in areas with severe topographic variations, often yield surprising and potentially useful results. If detailed surface elevation measurements are obtained over the grid, profiles can be corrected for topography; and archaeological features that would otherwise be difficult to discern in standard profiles may become apparent. The authors have used GPR at sites where large radar antennas (300 MHz in particular) had to be pulled over bumpy and hilly terrain. If the recorded reflection traces are stacked in order to average out the minor surface disturbances, and if background removal filters are applied to the data, good reflection data can still be obtained. Lower-frequency antennas (less than 300 MHz) are much less influenced by small changes in their roll and pitch when are pulled over rough surfaces, which is not the case when using higher-frequency antennas that are smaller in size. Even small surface irregularities that are not factored out by topographic corrections can severely disrupt subsurface reflections obtained from high-frequency antennas.

Potential users of GPR should not be discouraged by sites that have less than favorable ground surface conditions. One important factor in obtaining successful results is to make sure that lines within the surveyed grid are not distorted by the changing elevations along transects. The authors have found that it is imperative to take the time to square grid lines prior to conducting a survey in order to insure optimum subsurface accuracy. The spacing of lines within a grid must also be thought out in advance, taking into account the depth and dimensions of the targets and the frequency of the antenna to be used. This will insure that buried features of interest are completely illuminated and that none go undetected.

It is extremely important that field time be allotted for equipment calibration and velocity tests prior to conducting a survey. When arriving at a site, it is always tempting to begin acquiring reflection data immediately, before the field conditions are fully understood. This is especially true if the

GPR equipment is being rented by the day or hour! Haste of this sort many times yields unfortunate results, especially if it is discovered, after returning from the field, that important equipment adjustments such as trace stacking or time-window calibrations were made incorrectly. The same is true with time-depth conversions that can only be made accurately if velocity measurements are obtained as part of a field-data acquisition program.

Each GPR acquisition program will derive unique results depending on the field conditions at the time the survey was made. Soil and sediment moisture and other environmental factors will change with time, and radar-wave velocities and subsurface resolution will vary accordingly. If velocity tests are not made while in the field conducting the survey, they can never be accurately replicated; after-the-fact time-depth corrections will only be a matter of conjecture.

An understanding of the subsurface stratigraphy as it relates to reflections is also crucial in the interpretation process. This can only be accomplished if the stratigraphy in test excavations or outcrops is visible. If possible, some GPR profiles in a grid should be acquired that "tie" to known stratigraphy so that the reflections obtained in the profiles can be correlated to horizons of interest. If this is not done, the origin of individual reflections will always be in doubt, and some possibly important features may go unrecognized. In some situations, it may be difficult to get permission to excavate in order to observe and measure the stratigraphy, especially if the purpose of the survey is to map a site "noninvasively." If excavation is not allowed in an archaeologically sensitive area, it may still be possible to extend survey lines to nearby road cuts or to excavations located away from the site, but in a similar geological setting. As a last resort, it may be permissible to obtain stratigraphic information from auger probes or small-diameter cores that may be correlated to GPR profiles.

Users of GPR should always take into account information about soil and sediment conditions from published reports, or consult with geologists or soil scientists who have a familiarity with the area. If soils are very clay-rich, especially when wet, or if the features of interest are located quite deep in the ground, other geophysical techniques may be more useful than GPR. Although many other geophysical mapping methods can yield meaningful data, GPR is one of the few near-surface methods that can be calibrated accurately to give reliable depth information and produce three-dimensional maps.

During field data acquisition, it is important, but many times difficult, to refrain from making judgements about the success or failure of the survey. The typical question posed by onlookers is, "Have you found anything yet?" Although it is tempting to expound on the origin of anomalies that may be visible in raw data profiles on the computer screen or on paper printouts, processing and thoughtful interpretation of the data after returning from the field is usually necessary for any accurate assessment of a survey's success. The authors have often found that preliminary conclusions based on a perusal of raw profiles in the field are often inaccurate and any evaluation of the survey's success based on them can be both hasty and incorrect. After applying background removal filters, time-slice analyses, and other interpretative techniques to the data, otherwise invisible features often appear in the most unlikely places. A detailed analysis of the data is almost always necessary and casual approaches to data processing and interpretation will likely result in flawed or failed surveys.

For stratigraphically or archaeologically complicated sites, a dynamic process of computer manipulation and human cognitive interpretation is necessary. Individualized and sophisticated techniques must sometimes be improvised in order to deal with the complexities of some sites. This process may necessitate a cooperative analysis by a team of archaeologists, geologists, and geophysicists. In practice, this type of approach is often difficult due to economic and time constraints, necessitating that the archaeologist in charge of the survey be well-versed in many of the techniques discussed here.

This book was written with the knowledge that many of the processing and interpretation techniques presented will be superseded by more advanced data processing and higher resolution imaging in the near future. At present, many of the techniques discussed have only recently emerged from research and development to the application stage where archaeologists can employ them successfully. They will most likely be much improved upon soon as GPR becomes more widely employed.

GPR techniques use some of the same wave theory and field techniques that are commonly employed in seismic exploration by those in the petroleum business. Although the two methods use different energy sources (with petroleum exploration using seismic waves instead of radar waves), many of the same processing and interpretation techniques can be employed in both.

Near-surface GPR exploration has advanced to about the same stage as seismic exploration was in the 1970s, prior to the advent of much of our present computer technology. The borrowing of seismic acquisition and processing techniques holds the promise of advancing the GPR method rapidly into the twenty-first century. Some of the methods discussed in this book, such as slice-map generation, velocity analysis, and two-dimensional computer modeling, were modified directly from common petroleum-exploration programs.

Among the most promising potential advancements in the future will be the addition of multiple receiver/transmitter systems (van Deen 1996). In petroleum exploration, it is now common practice to use multiple receivers (sometimes many thousands), placed in an array on the ground surface, that can simultaneously "listen" for reflected energy that is transmitted from one or more surface stations. The reflected data are recorded on multiple computer channels and can then be processed in a three-dimensional fashion yielding superior results. To date, most GPR systems employ only single channels that both transmit and receive at one location on the ground surface. By increasing the number of receivers, images with much higher resolution will result. Recent work by Ulriksen (1992) suggests that a five-channel prototype GPR that is being used to test the integrity of roads in Sweden may have a future applicability in archaeological exploration. A commercial four-channel system has become available (GSSI SIR-10 system) that has yet to be implemented in a true three-dimensional array in archaeological mapping.

The vast amount of petroleum-exploration seismic software developed for multiple receivers could easily be adapted to multiple-receiver GPR systems. One must be cautious when using seismic processing programs "off the shelf," however, because there are important differences between seismic and radar wave propagation in the subsurface. Seismic waves tend to increase their velocity with depth, while radar waves slow down, usually due to increased water saturation with depth. This will create much different radar-wave paths because of subsurface changes in the refraction and transmission of waves across buried velocity interfaces. It is fortunate, however, that the similarities between the two methods far outweigh the differences.

With the inclusion of multiple receivers, the standard two-dimensional GPR velocity analyses discussed in this book could be expanded into three-dimensions to arrive at spacial velocity changes across a grid. The conversion of radar travel time to depth could then be made spatially, greatly increasing the accuracy of subsurface maps.

Most of the GPR work to date has focused on continuous radar wave propagation generated by a series of pulses at a surface antenna. New systems are being developed that will create radar "chirps," allowing radar energy of known measurable frequencies to be transmitted at various known pulse lengths. The reflections from these chirps could then be recorded in the same fashion as continuous records, but with a greater depth penetration and reflection resolution. Similar systems using seismic waves have been in use for many years with extraordinary results.

One of impediments to the technological advancement of GPR in archaeology is that the economic returns in most cases cannot justify high research and development investment. This has not been the case for seismic procedures developed for petroleum exploration where the returns for innovative technological development can be quite staggering! Nonetheless, GPR processing technology is steadily improving by borrowing ideas from the seismic method. New developments are also being made due to GPR's expanding role in engineering and environmental mapping. The search for hazardous waste sites, the quantitative measurement of road and airport runway integrity, and many other geotechnical applications are growing at exponential rates. These new applications being developed for commercial use will eventually become available to the archaeological community, dramatically expanding our ability to "see" below the ground surface.

References Cited

Annan, A. P., and L. T. Chua
    1992 Ground-Penetrating Radar Performance Predictions. In Ground Penetrating Radar, edited by J. A. Pilon, pp. 5–13. Geological Survey of Canada, Paper 90–4.

Annan, A. P., and S. W. Cosway
    1992 Simplified GPR Beam Model for Survey Design. Extended Abstract of 62nd Annual International Meeting of the Society of Exploration Geophysicists, New Orleans.

    1994 GPR Frequency Selection. Proceedings of the Fifth International Conference on Ground Penetrating Radar, pp. 747–760. Waterloo Centre for Groundwater Research, Waterloo, Canada.

Annan, A. P., and J. L. Davis
    1977 Impulse Radar Applied to Ice Thickness Measurements and Freshwater Bathymetry. Geological Survey of Canada, Report of Activities paper 77–1B:117–124.

    1992 Design and Development of a Digital Ground Penetrating Radar System. In Ground Penetrating Radar, edited by J. A. Pilon, pp. 49–55. Geological Survey of Canada, Paper 90–4.

Annan, A. P., W. M. Waller, D. W. Strangway, J. R. Rossiter, J. D. Redman, and R. D. Watts
    1975 The Electromagnetic Response of a Low-loss, 2-layer, Dielectric Earth for Horizontal Electric Dipole Excitation. Geophysics 40:285–298.

Archaeological Department, Shimane Prefecture
    1978 Guide to the Asuka Historical Museum. Kansai Press, Japan.

Arcone, S. A.
    1995 Numerical Studies of the Radiation Patterns or Resistivity Loaded Dipoles. Journal of Applied Geophysics 33:39–52.

Arita, K.
    1994 Nyutabaru Kofun. Unpublished research report no. 16, Shintomi Machi, Miyazaki Prefecture, Japan.

Arnold, J. E., E. L. Ambos, and D. O. Larson
    1997 Geophysical Surveys of Stratigraphically Complex Island California Sites: New Implications for Household Archaeology. Antiquity 71:157–168.

Balanis, C. A.
    1989 Advanced Engineering Electromagnetics. John Wiley and Sons, New York.

Basson, U., Y. Enzel, R. Amit, and Z. Ben-Avraham
    1994 Detecting and Mapping Recent Faults with a Ground-Penetrating Radar in the Alluvial Fans of the Arava Valley, Israel. Proceedings of the Fifth International Conference on Ground Penetrating Radar, pp. 777–788. Waterloo Centre for Groundwater Research, Waterloo, Canada.

Batey, R. A.
    1987 Subsurface Interface Radar at Sepphoris, Israel. Journal of Field Archaeology 14:1–8.

Beres, L., and H. Haeni
    1991 Application of Ground-Penetrating Radar Methods in
    Hydrogeologic Studies. Groundwater 29:375–386.

Bernabini, M., E. Brizzolari, L. Orlando, and G. Santellani
    1994 Application of Ground Penetrating Radar on Colosseum Pillars.
    Proceedings of the Fifth International Conference on Ground
    Penetrating Radar, pp. 547–558. Waterloo Centre for Groundwater
    Research, Waterloo, Canada.

Bevan, B. W.
    1977 Ground-Penetrating Radar at Valley Forge. Geophysical Survey
    Systems, North Salem, New Hampshire.

Bevan, B. W., and J. Kenyon
    1975 Ground-Penetrating Radar for Historical Archaeology. MASCA
    Newsletter 11(2):2–7.

Bjelm, L.
    1980 Geologic Interpretation of SIR Data from a Peat Deposit in
    Northern Sweden. Unpublished manuscript, Department of Engi-
    neering Geology, Lund Institute of Technology, Lund, Sweden.

Bucker, F., M. Gurtner, H. Hortsmeyer, A. G. Green, and P. Huggenberger
    1996 Three-Dimensional Mapping of Glaciofluvial and Deltaic
    Sediments in Central Switzerland Using Ground Penetrating Radar.
    Proceedings of the Sixth International Conference on Ground
    Penetrating Radar, pp. 45–50. Department of Geoscience and
    Technology, Tohoku University, Sendai, Japan.

Butler, D. K., J. E. Simms, and D. S. Cook
    1994 Archaeological Geophysics Investigation of the Wright Brothers
    1910 Hanger Site. Geoarchaeology: An International Journal
    9:437–466.

Cai, J., and G. A. McMechan
    1994 Ray-Based Synthesis of Bistatic Ground Penetrating Radar
        Profiles. Proceedings of the Fifth International Conference on
        Ground Penetrating Radar, pp. 19–29. Waterloo Centre for
        Groundwater Research, Waterloo, Canada.

Clark, A.
    1996 Seeing beneath the Soil: Prospecting Methods in Archaeology. B.
        T. Batsford, London.

Collins, M. E.
    1992 Soil Taxonomy: A Useful Guide for the Application of Ground-
        Penetrating Radar. Fourth International Conference on Ground-
        Penetrating Radar. Edited by P. Hanninen and S. Autio, pp. 125–
        132. Geological Survey of Finland Special Paper 16, Rovaniemi,
        Finland.

Conyers, L. B.
    1995a The Use of Ground-Penetrating Radar to Map the Buried
        Structures and Landscape of the Ceren Site, El Salvador.
        Geoarchaeology 10:275–299.

    1995b The Use of Ground-Penetrating Radar to Map the Topography
        and Structures at the Ceren Site, El Salvador. Unpublished Ph.D.
        dissertation, Department of Anthropology, University of Colorado,
        Boulder.

Conyers, L. B., and J. E. Lucius
    1996 Velocity Analysis in Archaeological Ground-Penetrating Radar
        Studies. Archaeological Prospection 3:25–38.

Cook, J. C.
    1973 Radar Exploration Through Rock in Advance of Mining. Trans-
        actions of the Society of Mineral Engineering AIME 254:140–146.

    1975 Radar Transparencies of Mines and Tunnel Rocks. Geophysics
        40:865–885.

Czarnowski, J., S. Geibler, and A. F. Kathage
1996 Combined Investigation of GPR and High Precision Real-Time Differential GPS. Proceedings of the Sixth International Conference on Ground Penetrating Radar, pp. 207–209. Department of Geoscience and Technology, Tohoku University, Sendai, Japan.

Davis, J. L. and A. P. Annan
1989 Ground-Penetrating Radar for High-Resolution Mapping of Soil and Rock Stratigraphy. Geophysics 37:531–551.

1992 Applications of Ground Penetrating Radar to Mining, Groundwater, and Geotechnical Projects: Selected Case Histories. In Ground Penetrating Radar, edited by J. A. Pilon, pp. 49–56. Geological Survey of Canada Paper 90-4, Ottawa.

Deng, S., Z. Zhengrong, and H. Wang
1994 The Application of Ground Penetrating Radar to Detection of Shallow Faults and Caves. Proceedings of the Fifth International Conference on Ground Penetrating Radar, pp. 1115–1133. Waterloo Centre for Groundwater Research, Waterloo, Canada.

De Vore, S. L.
1990 Ground-Penetrating Radar as a Survey Tool in Archaeological Investigations: An Example from Fort Laramie National Historic Site. The Wyoming Archaeologist 33:23–38.

Dobrin, M. B.
1976 Introduction to Geophysical Prospecting. McGraw-Hill, New York.

Dolphin, L. T., R. L. Bollen, and G. N. Oetzel
1974 An Underground Electromagnetic Sounder Experiment. Geophysics 39:49–55.

Doolittle, J. A.
1982 Characterizing Soil Map Units with the Ground-Penetrating Radar. Soil Survey Horizons 23:3–10.

Doolittle, J. A., and L. E. Asmussen
  1992 Ten Years of Applications of Ground Penetrating Radar by the
      United States Department of Agriculture. In Fourth International
      Conference on Ground-Penetrating Radar, edited by P. Hanninen
      and S. Autio, pp. 139–147. Geological Survey of Finland Special
      Paper 16. Rovaniemi, Finland

Doolittle, J. A., and W. F. Miller
  1991 Use of Ground-Penetrating Radar Techniques in Archaeological
      Investigations. In Applications of Space-Age Technology in
      Anthropology Conference Proceedings, Second Edition. NASA
      Science and Technology Laboratory, Stennis Space Center, Missis-
      sippi.

  1992 Geophysical Investigations at the Ceren Site, El Salvador. In
      1992 Investigations at the Ceren Site, El Salvador: A Preliminary
      Report, edited by P. D. Sheets, and K. A. Kievit, pp. 10–19.
      Department of Anthropology, University of Colorado, Boulder.

Duke, S.
  1990 Calibration of Ground-Penetrating Radar and Calculation of
      Attenuation and Dielectric Permittivity Versus Depth. Unpublished
      master's thesis, Department of Geophysics, Colorado School of
      Mines, Golden, Colorado.

Engheta, N., C. H. Papas, and C. Elachi
  1982 Radiation Patterns of Interfacial Dipole Antennas. Radio Science
      17:1557–1566.

Fenner, T. J.
  1992 Recent Advances in Subsurface Interface Radar Technology. In
      Fourth International Conference on Ground-Penetrating Radar,
      edited by P. Hanninen and S. Autio, pp. 13–19. Geological Survey
      of Finland Special Paper 16. Rovaniemi, Finland

Fischer, P. M., S. G. W. Follin, and P. Ulriksen
  1980 Subsurface Interface Radar Survey at Hala Sultan Tekke,
    Cyprus. In Applications of Technical Devices in Archaeology,
    edited by P. M. Fischer. Studies in Mediterranean Archaeology
    63:48–51.

Fisher, E., G. A. McMechan, and A. P. Annan
  1992 Acquisition and Processing of Wide-Aperture Ground-Penetrat-
    ing Radar Data. Geophysics 57:495–504.

Fisher, S. C., R. R. Stewart, and H. M. Jol
  1994 Processing Ground Penetrating Radar Data. In Proceedings of
    the Fifth International Conference on Ground Penetrating Radar,
    pp. 661–675. Waterloo Centre for Groundwater Research, Water-
    loo, Canada.

Fullagar, P. K., and D. Livleybrooks
  1994 Trial of Tunnel Radar for Cavity and Ore Detection in the
    Sudbury Mining Camp, Ontario. In Proceedings of the Fifth
    International Conference on Ground Penetrating Radar, pp. 883–
    894. Waterloo Centre for Groundwater Research, Waterloo,
    Canada.

Geophysical Survey Systems, Inc.
  1987 Operations Manual for Subsurface Interface Radar System-3.
    Manual #MN83-728. Geophysical Survey Systems, North Salem,
    New Hampshire.

Goodman, D.
  1994 Ground-Penetrating Radar Simulation in Engineering and
    Archaeology. Geophysics 59:224–232.

  1996 Comparison of GPR Time Slices and Archaeological Excava-
    tions. Proceedings of the Sixth International Conference on Ground
    Penetrating Radar, pp. 77-82. Department of Geoscience and
    Technology, Tohoku University, Sendai, Japan.

Goodman, D., and Y. Nishimura
1993 A Ground-Radar View of Japanese Burial Mounds. Antiquity
67:349–354.

Goodman, D., Y. Nishimura, and J. D. Rogers
1995 GPR Time-Slices in Archaeological Prospection. Archaeological
Prospection 2:85–89.

Goodman, D., Y. Nishimura, R. Uno, and T. Yamamoto
1994 A Ground Radar Survey of Medieval Kiln Sites in Suzu City,
Western Japan. Archaeometry 36:317–326.

Grasmueck, M.
1994 Application of Seismic Processing Techniques to Discontinuity
Mapping with Ground-Penetrating Radar in Crystalline Rock of
the Gotthard Massif, Switzerland. Proceedings of the Fifth International Conference on Ground Penetrating Radar, pp. 1135–1139.
Waterloo Centre for Groundwater Research, Waterloo, Canada.

Hanninen, P., P. Hanninen, L. Koponen, A. Koskaihde, P. Maijala, R.
Pollari, T. Saarenketo, and A. Sutinen
1992 Ground Penetrating Radar. The Finnish Geotechnical Society
and the Finnish Building Centre, Tammer-Paino, Oy, Finland.

Hiroshi, T.
1989 Kodaisi Fukugen. Koudansha, Tokyo.

Huggenberger, P., E. Meier, and M. Beres
1994 Three-Dimensional Geometry of Fluvial Gravel Deposits from
GPR Reflection Patterns; a Comparison of Results of Three
Different Antenna Frequencies. Proceedings of the Fifth Interna-
tional Conference on Ground Penetrating Radar, pp. 805–815.
Waterloo Centre for Groundwater Research, Waterloo, Canada.

Imai, T., T. Sakayama, and T. Kanemori
1987 Use of Ground-Probing Radar and Resistivity Surveys for
Archaeological Investigations. Geophysics 52:137–150.

Isseki, N.
  1993 Unpublished research report: Asada Site, Shibukawa Shi.
  Nakasuji Isseki Report, Guma Prefecture, Japan.

Jackson, J. D.
  1977 Classical Electrodynamics. John Wiley, New York.

Johnson, R. W., R. Glaccum, and R. Wotasinski
  1980 Application of Ground Penetrating Radar to Soil Survey. Soil
  Crop Science Society Proceedings 39:68–72.

Jol, H. M., and Smith, D. G.
  1992 Ground Penetrating Radar of Northern Lacustrine Deltas.
  Canadian Journal of Earth Science 28:1939–1947.

Keller, G. V.
  1988 Rock and Mineral Properties. In Applications in Electromagnetic
  Methods in Applied Geophysics, vol. 1, edited by M. N.
  Nabighian, pp.13–24. Society of Exploration Geophysics, Tulsa,
  Oklahoma.

Kemerait, R. C.
  1994 Ground Penetrating Radar Considerations for Optimizing the
  Data Collection Scenario. Proceedings of the Fifth International
  Conference on Ground Penetrating Radar, pp. 761–775. Waterloo
  Centre for Groundwater Research, Waterloo, Canada.

Kenyon, J. L.
  1977 Ground-Penetrating Radar and Its Application to a Historical
  Archaeological Site. Historical Archaeology 11:48–55.

Kievit, K. A.
  1994 Jewel of Ceren: Form and Function Comparisons for the
  Earthern Structures of Joya de Ceren, El Salvador. Ancient
  Mesoamerica 5:193–208.

Kobayashi, T., T. Yamada, and Y. Tanaka
    1993 Obatake Kiln Site, Suzu City. Unpublished magnetic report, Department of Earth Sciences, Ishikawa Ken, Toyama University, pp. 67–75.

Kraus, J. D.
    1950 Antennas. McGraw-Hill, New York.

Kutrubes, D. L.
    1986 Dielectric Permittivity Measurements of Soils Saturated with Hazardous Fluids. Unpublished master's thesis, Department of Geophysics, Colorado School of Mines, Golden.

LaFlech, P. T., J. P. Todoeschuck, O. G. Jensen, and A. S. Judge
    1991 Analysis of Ground-Penetrating Radar Data: Predictive Deconvolution. Canadian Geotechnical Journal 28:134–139.

Lanz, E., L. Jemi, R. Muller, A. Green, A. Pugin, and P. Huggenberger
    1994 Integrated Studies of Swiss Waste Disposal Sites: Results from Georadar and other Geophysical Surveys. Proceedings of the Fifth International Conference on Ground Penetrating Radar, pp. 1261–1274. Waterloo Centre for Groundwater Research, Waterloo, Canada.

Lehmann, F., H. Hortsmeyer, A. Green, and J. Sexton
    1996 Georadar Data from the Northern Sahara Desert: Problems and Processing Strategies. Proceedings of the Sixth International Conference on Ground Penetrating Radar, pp. 51–56. Department of Geoscience and Technology, Tohoku University, Sendai, Japan.

Loker, W. M.
    1983 Recent Geophysical Explorations at Ceren. In Archaeology and Volcanism in Central America, edited by P. D. Sheets, pp. 254–274. University of Texas Press, Austin.

Machida, H., and F. Arai
    1983 Extensive Ash Falls in and around the Sea of Japan from Large
          Late Quaternary Eruption. Journal of Volcanology and Geothermal
          Research 18:151–164.

Maijala, P.
    1992 Application of Some Seismic Data Processing Methods to
          Ground Penetrating Radar Data. In Fourth International Confer-
          ence on Ground-Penetrating Radar, edited by P. Hanninen and S.
          Autio, pp. 103–110. Geological Survey of Finland Special Paper
          16. Rovaniemi, Finland

Malagodi, S., L. Orlando, and S. Piro
    1994 Improvement of Signal to Noise Ratio of Ground Penetrating
          Radar Using CMP Acquisition and Data Processing. Proceedings
          of the Fifth International Conference on Ground Penetrating Radar,
          pp. 689–699. Waterloo Centre for Groundwater Research, Water-
          loo, Canada.

    1996 Approaches to Increase Resolution of Radar Signal. Proceedings
          of the Sixth International Conference on Ground Penetrating
          Radar, pp. 283–288. Department of Geoscience and Technology,
          Tohoku University, Sendai, Japan.

Malagodi, S., L. Orlando, and F. Rosso
    1996 Location of Archaeological Structures Using GPR Method:
          Three-Dimensional Data Acquisition and Radar Signal Processing.
          Archaeological Prospection 3:13–23.

Miller, C. D.
    1989 Stratigraphy of Volcanic Deposits at El Ceren. In Preliminary
          Report: Ceren Project 1989, pp. 8–19. Department of Anthropol-
          ogy, University of Colorado, Boulder.

Milligan, R., and M. Atkin
    1993 The Use of Ground-Probing Radar within a Digital Environment
    on Archaeological Sites. In Computing the Past: Computer Appli-
    cations and Quantitative Methods in Archaeology, edited by J.
    Andresen, T. Madsen, and I. Scollar, pp. 21–33. Aarhus University
    Press, Aarhus, Denmark.

Moffat, D. L., and R. J. Puskar
    1976 A Subsurface Electromagnetic Pulse Radar. Geophysics 41:506–
    518.

Neves, F. A., J. A. Miller, and M. S. Roulston
    1996 Source Signature Deconvolution of Ground Penetrating Radar
    Data. Proceedings of the Sixth International Conference on Ground
    Penetrating Radar, pp. 573–578. Department of Geoscience and
    Technology, Tohoku University, Sendai, Japan.

Noon, D. A., D. Longstaff, and R. J. Yelf
    1994 Advances in the Development of Step Frequency Ground
    Penetrating Radar. Proceedings of the Fifth International Confer-
    ence on Ground Penetrating Radar, pp. 117–131. Waterloo Centre
    for Groundwater Research, Waterloo, Ontario, Canada.

Olhoeft, G. R.
    1981 Electrical Properties of Rocks. In Physical Properties of Rocks
    and Minerals, edited by Y. S. Touloukian, W. R. Judd, and R. F.
    Roy, pp. 257–330. McGraw-Hill, New York.

    1986 Electrical Properties from 10^{-3} to 10^9 Hz-Physics and Chemistry.
    Proceedings of the 2nd International Symposium of Physics and
    Chemistry of Porous Media, edited by J. R. Bananvar, J. Koplik,
    and K. W. Winkler, pp. 281–298. Schlumberger-Doll, Ridgefield,
    Connecticut.

    1994a Modeling Out-of-Plane Scattering Effects. Proceedings of the
    Fifth International Conference on Ground Penetrating Radar, pp.
    133–144. Waterloo Centre for Groundwater Research, Waterloo,
    Ontario, Canada.

1994b Geophysical Observations of Geological, Hydrological and
Geochemical Heterogeneity. Proceedings of the Symposium on the
Application of Geophysics to Engineering and Environmental
Problems, pp. 129–141. Boston.

Olhoeft, G. R., and D. E. Capron
1993 Laboratory Measurements of the Radio Frequency Electrical and
Magnetic Properties of Soils from near Yuma, Arizona. U. S.
Geological Survey Open File Report 93–701.

Olson, C. G., and Doolittle, J. A.
1985 Geophysical Techniques for Reconnaissance Investigation of
Soils and Surficial Deposits in Mountainous Terrain. Soil Science
Society of America Journal 49:1490–1498.

Powers, M. H., and G. R. Olhoeft
1994 GPRMODV2: One-Dimensional Full Waveform Forward
Modeling of Dispersive Ground Penetrating Radar Data. U. S.
Geological Survey Open File Report 95–58.

1995 Waveform Forward Modeling of Dispersive Ground Penetrating
Radar. U.S. Geological Survey Open File Report 95–58.

Rees, H. V., and J. M. Glover
1992 Digital Enhancement of Ground Probing Radar Data. In Ground
Penetrating Radar, edited by J. A. Pilon, pp. 187–192. Geological
Survey of Canada Paper 90–4.

Rogers, J. D., D. G. Wycoff, and D. A. Peterson (editors)
1989 Contributions to Spiro Archaeology: Mound Excavations and
Regional Perspectives. Oklahoma Archaeological Survey, Studies
in Oklahoma's Past 16.

Sakai, H., T. Koyayashi, T. Yamada, and Y. Tanaka
1993 Magnetic Survey of the Obatake Kiln Site in Suzu City. Suzu
Obatake Kiln Site Report. Toyama University, Ishikawa Ken, pp.
67-75.

Scollar, I., A. Tabbagh, A. Hesse, and I. Herzog
1990 Archaeological Prospecting and Remote Sensing. Cambridge University Press, Cambridge, England.

Sellman, P. V., S. A. Arcone, and A. J. Delaney
1983 Radar Profiling of Buried Reflectors and the Ground Water Table. Cold Regions Research and Engineering Laboratory Report 83–11:1–10.

Sheets, P. D.
1992 The Ceren Site: A Prehistoric Village Buried by Volcanic Ash in Central America. Harcourt Brace Jovanovich, Fort Worth.

Sheets, P. D., W. M. Loker, H. A. W. Spetzler, and R. W. Ware
1985 Geophysical Exploration for Ancient Maya Housing at Ceren, El Salvador. National Geographic Research Reports 20:645–656.

Sheriff, R. E.
1984 Encyclopedic Dictionary of Exploration Geophysics. Second Edition. Society of Exploration Geophysics, Tulsa, Oklahoma.

Shih, S. F., and J. A. Doolittle
1984 Using Radar to Investigate Organic Soil Thickness in the Florida Everglades. Soil Science Society of America Journal 48:651–656.

Smith, D. G., and H. M. Jol
1995 Ground Penetrating Radar: Antenna Frequencies and Maximum Probable Depths of Penetration in Quaternary Sediments. Journal of Applied Geophysics 33:93–100.

Sternberg, B. K., and J. W. McGill
1995 Archaeology Studies in Southern Arizona Using Ground Penetrating Radar. Journal of Applied Geophysics 33:209–225.

Stove, G. C., and P. V. Addyman
1989 Ground Probing Impulse Radar: An Experiment in Archaeological Remote Sensing at York. Antiquity 63:337–342.

Sun, J., and R. A. Young.
   1995 Scattering in Ground-Penetrating Radar Data. Geophysics
      6:1378–1385.

Tillard, S., and J. Dubois
   1995 Analysis of GPR Data: Wave Propagation Velocity Determina-
      tion. Journal of Applied Geophysics 33:77–91.

Todoeschuck. J. P., P. T. LaFleche, O. G. Jensen, A. S. Judge, and
J. A. Pilon
   1992 Deconvolution of Ground Probing Radar Data. In Ground
      Penetrating Radar, edited by J. A. Pilon, pp. 227–230. Geological
      Survey of Canada Paper 90-4.

Tyson, P.
   1994 Noninvasive Excavation. Technology Review, February/March
      1994:20–21.

Uno, T., K. Maekawa, K. Suzuki, S. Hamaki, and K. Miyazawa
   1993 Suzu Obatake Kilns: A Report of the Excavation of Medieval
      Sue Pottery Kilns in Ishikawa, Japan. Toyama University.

Urliksen, C. P. F.
   1992 Multistatic Radar System-MRS. In Fourth International Confer-
      ence on Ground-Penetrating Radar, edited by P. Hanninen and S.
      Autio, pp. 57–63. Geological Survey of Finland Special Paper 16.
      Rovaniemi, Finland

van Deen, J. K.
   1996 3-D Ground Probing Radar. Proceedings of the Sixth Interna-
      tional Conference on Ground Penetrating Radar, pp. 295–298.
      Department of Geoscience and Technology, Tohoku University,
      Sendai, Japan.

van Heteren, S., Fitzgerald, D. M., and McKinlay, P. S.
    1994 Application of Ground-Penetrating Radar in Coastal Strati-
    graphic Studies. Proceedings of the Fifth International Conference
    on Ground Penetrating Radar, pp. 869–881. Waterloo Centre for
    Groundwater Research, Waterloo, Canada.

van Overmeeren, R. A.
    1994 High Speed Georadar Data Acquisition for Groundwater Explo-
    ration in the Netherlands. Proceedings of the Fifth International
    Conference on Ground Penetrating Radar, pp. 1057–1073. Water-
    loo Centre for Groundwater Research, Waterloo, Canada.

Vaughan, C. J.
    1986 Ground-Penetrating Radar Surveys Used in Archaeological
    Investigations. Geophysics 51:595–604.

Vickers, R. S., and L. T. Dolphin
    1975 A Communication on an Archaeological Radar Experiment at
    Chaco Canyon, New Mexico. MASCA Newsletter 11(1).

Vickers, R. S., L. T. Dolphin, and D. Johnson
    1976 Archaeological Investigations at Chaco Canyon Using Subsur-
    face Radar. In Remote Sensing Experiments in Cultural Resource
    Studies, edited by T. R. Lyons, pp. 81–101. Chaco Center, USDI-
    NPS and the University of New Mexico, Albuquerque.

von Hippel, A. R.
    1954 Dielectrics and Waves. MIT Press, Cambridge, Massachusetts.

Walden, A. T., and J. W. J. Hosken
    1985 An Investigation of the Spectral Properties of Primary Reflection
    Coefficients. Geophysical Prospecting 33: 400–435.

Wright, D. C., G. R. Olhoeft, and R. D. Watts
    1984 GPR Studies on Cape Cod. Proceedings of the National Water
    Well Association Conference on Surface and Borehole Geophysi-
    cal Methods, pp. 666–680. San Antonio, Texas.

Young, R. A., and S. Jingsheng
　1994　Recognition and Removal of Subsurface Scattering in GPR Data. Proceedings of the Fifth International Conference on Ground Penetrating Radar, pp. 735–746. Waterloo Centre for Groundwater Research, Waterloo, Canada.

Yu, H., Y. Xiaojian, and S. Yuansheng
　1996　The Use of Fk Techniques in GPR Processing. In Proceedings of the Sixth International Conference on Ground Penetrating Radar, pp. 595–600. Department of Geoscience and Technology, Tohoku University, Sendai, Japan.

Index

A

absorption of energy, 55
adjustments
   equipment, 68
   software, 68
agricultural run-off, 35
air wave, 122
amplitude
   variations, 152, 158
   reflections, 27
amplitude analysis, 149–50
amplitude anomaly maps, 173, 190, 192
ancient landscape, 145
ancient living surface, 104
anomaly
   amplitude, 150–1
   depth, 15
   hunting, 196
antenna
   continuous movement, 25–6
   dipole, 35, 58
   frequency constraints, 40–1, 46, 64, 200
   movement, 61, 66
   step movement, 25, 61, 67
   position, 59
   selection, 51–2
   shielding, 37–8
   sizes, 43–4
   speed, 24
Asada site, Japan, 172, 174, 176–7
attenuation, 28, 53

B

backfilled excavations, 192
background removal, 78, 203
band width, 23, 40
bar test, 109
bedding planes, 18
bulk density, 32
burial chambers, 165–6, 185, 199
burial mounds, 20
burial sequences, 162
burials, 91, 199
buried features
   pipes, 62
   structures, 94
   walls, 19

buried surfaces, resolution, 48
buried topography, 98

C

cable operator, 67
cables, 58
caliche soils, 35
center frequency, 42, 45, 55
Ceren site, El Salvador, 18, 49, 79, 94,
        98–99, 109, 115, 131, 137, 139–
        40, 143, 146, 150
Chaco Canyon, New Mexico, 18
clay, effect on attenuation, 46
clay artifacts, 199
clay floors, 79, 105, 141
clay layers, 94, 200
clay soil, 53, 55
clay content, 163, 169
clutter, 50
coefficient of reflectivity, 34
columns, 105, 142
common-mid point test, 119, 126
conductivity, of media, 14, 35
cone of transmission, 38–9
construction material, 94
contacts, lithologic, 18
control unit, 58
coupling, 55
cultural layers, 20

D

data
    analog, 59
    band-pass filtering, 74
    digitization, 58
    filtering, 74

graphic recording, 31
horizontal filtering, 71
magnetic tape recording, 31
processing, 77
raw, 150
recording, 60
smoothing, 73
stacking, 70–1
vertical filtering 74
data storage
    analog, 26
    digital, 28
deconvolution, 81
depth of penetration, 31
depth slice, 151
dielectric materials, 32
dielectric permeability, 32
diffraction, 30
digitization, 60
dispersion, 28, 38
dog leg path, 89–90
downloading, 41
drainage patterns, 142, 144

E

Edo Period, Japan, 160
electrical cables, 62
electrical conductivity, 32, 34
electromagnetic energy, 34
electromagnetic field, 14
    orientation of, 121
electromagnetic induction surveys, 14
electron displacement, 14
equipment, manufacturers, 57
excavation trenches, 197

F

fences, 64
fiber-optics, 58
filtering
  deconvolution, 81
  Fk, 80
  migration, 82
  background removal, 78
fire pits, 197
f-k filters, 80
FM interference, 51, 73, 75
focusing, 38, 52–53
footprint, 37, 50–51
  measuring, 39
Fort Laramie Historic Site, Wyoming,
  20
forward modeling, 86
foundations, 170, 197
frequency
  distribution, 41–2
  radar energy, 23
  relation to wavelength, 45, 47
frozen lakes and rivers, 26

G

garbage pits, 181
global positioning systems (GPS), 63
grid
  construction, 65, 161, 201
  grid, dimensions, 25, 62
  non-rectangular, 63
  orientation, 62
ground, level, 201
ground coupling, 35
groundwater, depth to, 18

H

Hala Sultan Teke site, Cyprus, 19
hard drive, storage capacity, 70
header information, 68
horizontal scale, corrections, 61
horizon slice surface, 175
horizon slices, 172
house floors, 20, 105
house platforms, 19–20, 96

I

illumination, 39
images
  animated, 145
  three-dimensional, 145
  virtual, 145
interpolation, radius of, 155

J

joint systems, 18

K

keyhole mound, 167
kilns, 178–9, 182, 184, 187, 198
Kofun 102, 103 mounds, Japan, 104,
  185, 188–9, 191
Kofun 111 mound, Japan, 85, 99, 163–4
Kualoa site, Oahu, Hawaii, 20
Kyushu, Japan, volcanoes, 161

L

laboratory measurements, 130
living surfaces, 92, 95, 137, 198

M

magnetic anomaly, 181, 184
magnetic gradiometer, 180
magnetic permeability, 32, 34
magnetic susceptibility surveys, 14–5
markers, hand held, 60
matrix, contrasts, 27
Matsuzaki site, Japan, 103, 178, 182–3
metal
    reflectivity of, 35
    objects, 109, 133, 199
metallic tools, 199
migration, 82
Mississippian Period, 190
moats, 53–4, 91, 197
models
    creation of, 86
    forward, 86
    synthetic, 83
    two dimensional, 28
moisture, 32, 132–3, 202
molecular relaxation, 46
mounds, burial, 163
multiples, 84, 91–2
multiple channel systems, 204

N

Nanaojo Castle, Japan, 154–5
near-field zone, 55–6, 139, 180, 189
noise, 52
Nyutabaru site, Japan, 18, 156–7, 159, 162

O

overhead power lines, 64

P

penetration depth, 23
pit dwellings, 172–3, 179, 181–2, 197
planar surfaces, resolution, 48
plowing, 161, 168, 179
point source reflection, 30
point sources/targets, 50, 82, 84
porosity, 32
pottery sherds, 199
profile, 29
    raw data, 15
propagation, radar energy, 35
pull-down, 93
pull-up, 93
pulse, signature, 42–3

R

radar energy
    leakage, 124
    penetration, 163
radar waves, velocity of, 27
radargram
    creation of, 86
    synthetic, 83
radiation pattern, 38
range gains, 73–4
raw data profiles, 15
ray paths, 87, 89
ray tracing, 86
RDP, determination of 114, 129, 131
recording rate, 70
Red Bay, Labrador, 19
reflections
    creation of, 26–7, 30, 35
    collapse of, 78
    multiples, 84

reflectivity, 36–7
refraction, 35
relative dielectric permeability (RDP), 32–3, 114, 129, 131
resistivity map, 180, 182
resistivity surveys, 14
ringing, 58, 78
Rockwell Mount site, Illinois, 20

S

Saitobaru Park, Japan, 163, 185
saltwater, 16, 35
samples per scan, 69
sampling, incremental, 69
sample rate, 70
scale
    horizontal, 61
    vertical, 62
scattering, 52–3
sediment
    dry, 16
    wet, 16
seismic data, processing, 77, 204
shadow zone, 94
shafts, 185
Shawnee Creek site, Missouri, 18, 103, 190, 192–3
signal position, 72
slice map, 151
    construction of, 153–4
    level ground, 156
    sloping surfaces, 48
Snell's Law, 36
soil units, 87
    buried, 18
speed of light (C), 33

Spiro Mounds site, Oklahoma, 102, 168, 170
stacking, 70–1
standing walls, 94
stone tools, 198
stratigraphic correlation test, 115
stratigraphic layers, 200
stratigraphy, resolution of, 70
structures, 141, 144, 198
subsurface coverage, 29
surface vegetation, 80
Suzu City, Japan, 18, 184, 186, 187

T

target dimensions, 201
TBJ reflector, 138, 142
three-dimensional mapping, 145, 165
three-dimensional objects, resolution of, 46
time, conversion to depth, 108
time slices, horizon, 172
time slices, horizontal, 169, 179, 185, 192
time window, 68, 72
time-depth analysis, 107, 203
time-depth corrections, 202
time-slice analysis, 151, 203
time-slice maps, 151
topographic corrections, 62, 164, 168, 170–1, 185
trace, 29
transducer, 59
transects
    in grid, 62
    random, 62
    sinuous, 67

transillumination tests, 119–20
transmission rate, 29, 71
trees, 64
trenches, 91–2, 97, 197
tunnels, 50, 62, 85, 197
two-way time, 26, 61, 107, 152

V

variable soils, 80
velocity
   changes across site, 109
   changes with seasons, 109
   determination of, 112
   direct wave methods, 108–9
velocity analysis, 151, 158
velocity determination, 108
velocity of materials, 32–3
velocity pull-down, 93
velocity pull-up, 93
velocity tests, 19
velocity, reflected wave method, 108–9

vertical features, 184–85
vertical scale, corrections, 62
void spaces, 23, 50
volcanic beds, 141
volcanic soils, 80

W

wall test, 109, 112
walls, 142, 170, 200
water saturation, 178
water table, 178
wavelength, 47
   changes in ground, 46
   relation to frequency, 45, 47
   resolution, 48–9
wiggle trace, 77

Y

Yamashiro Futagozuka mound, Japan,
   101, 166
York, England, 20

Index of Authors Cited

Annan and Chua, 83
Annan and Cosway, 35–8, 40, 46
Annan and Davis, 26, 31, 58
Annan et al., 32, 36
Archaeological Department, Shimane
    Prefecture, 186
Arcone, 35
Arita, 162
Arnold et al., 151

Balanis, 55
Basson et al., 18
Batey, 31, 53
Beres and Haeni, 18
Bernabini et al., 119
Bevan, 25
Bevan and Kenyon, 19
Bjelm, 18
Bucker et al., 74
Butler et al., 21

Cai and McMechan, 83
Clark, 14
Collins, 18
Conyers, 21, 25, 94, 106, 116, 137
Conyers and Lucius, 109, 115
Cook, 18

Czarnowski et al., 63

Davis and Annan, 12, 26, 33, 35, 47,
    58, 70
Deng et al., 18
De Vore, 20
Dobrin, 32
Dolphin, 18
Doolittle, 18
Doolittle and Asmussen, 18, 26
Doolittle and Miller, 20, 25, 46, 53, 62,
    109
Duke, 46, 53

Engheta et al., 35, 41, 55

Fenner, 40, 74
Fisher et al., 19, 31, 55, 70–1, 73, 75,
    81–2, 119, 127
Fullagar and Livleybrooks, 18

Geophysical Survey Systems, Inc., 31–
    3, 73
Goodman, 20, 36, 83–5, 90, 93, 193
Goodman and Nishimura, 20, 83, 85
Goodman et al., 20–1, 94, 184, 186
Grasmueck, 70–1, 82

Hanninen et al., 119
Hiroshi, 156, 178
Huggenberger et al., 40

Imai et al., 19, 21, 119
Isseki, 172

Jackson, 90
Johnson et al., 18
Jol and Smith, 18, 26

Keller, 46
Kemerait, 68
Kenyon, 19
Kievit, 94, 142, 144
Kobayashi et al., 186
Kraus, 58
Kutrubes, 131

LaFlech et al., 81
Lanz et al., 37
Lehmann et al., 80
Loker, 31, 84, 138

Machida and Arai, 157
Maijala, 70–1, 73, 80–2, 119
Malagodi et al., 40, 77, 81–2, 119, 151
Miller, 115, 117, 133
Milligan and Atkin, 28, 77–8, 81–2,
    119, 137, 151
Moffat and Puskar, 18

Neves et al., 81
Noon et al., 35

Olhoeft, 27, 32, 46, 131–3
Olhoeft and Capron, 131
Olson and Doolittle, 18, 40

Powers and Olhoeft, 42, 86, 122

Rees and Glover, 81
Rogers et al., 168

Sakai et al., 184, 186
Scollar et al., 180, 182
Sellman et al., 27, 34
Sheets, 94, 115, 144
Sheets et al., 19, 138
Sheriff, 34, 36, 55
Shih and Doolittle, 18, 46, 53, 61, 73,
    78
Smith and Jol, 40
Sternberg and McGill, 21, 58, 73, 78,
    108, 133
Stove and Addyman, 20
Sun and Young, 63

Tillard and Dubois, 119
Todoeschuck et al., 81
Tyson, 21

Uno et al., 184
Urliksen, 204

van Deen, 204
van Heteren et al., 18, 46
van Overmeeren, 18
Vaughan, 19, 21
Vickers and Dolphin, 16
Vickers et al., 18
von Hippel, 32

Walden and Hosken, 34
Wright et al., 26

Young and Jingsheng, 82
Yu et al., 81